Conference Board of the Mathematical Sciences

REGIONAL CONFERENCE SERIES IN MATHEMATICS

supported by the

National Science Foundation

Number 26

CLASS GROUPS AND PICARD GROUPS OF GROUP RINGS AND ORDERS

by

IRVING REINER

Published for the
Conference Board of the Mathematical Sciences
by the
American Mathematical Society
Providence, Rhode Island

Expository Lectures
from the CBMS Regional Conference
held at Carleton College
August 11–15, 1975

AMS (MOS) 1970 subject classifications. Primary 16-02, 16A18, 16A54, 20C10;
Secondary 13D15, 18-02, 20-02

Library of Congress Cataloging in Publication Data

Reiner, Irving.
Class groups and Picard groups of group rings and orders.

(Regional conference series in mathematics ; no. 26)
"Expository lectures from the CBMS regional conference held at Carleton College, August 11-15, 1975."
Includes bibliographical references and index.
1. Class groups (Mathematics) 2. Picard groups. 3. Fields, Algebraic. 4. Ideals (Algebra) 5. Group rings. I. Conference Board of the Mathematical Sciences. II. Title. III. Series.
QA1.R33 no. 26 [QA247] 510'.8s [512'.2] 76-10337
ISBN 0-8218-1676-4

CONTENTS

Other Monographs in this Series

No. 1. Irving Kaplansky: *Algebraic and analytic aspects of operator algebras*
2. Gilbert Baumslag: *Lecture notes on nilpotent groups*
3. Lawrence Markus: *Lectures in differentiable dynamics*
4. H.S.M. Coxeter: *Twisted honeycombs*
5. George W. Whitehead: *Recent advances in homotopy theory*
6. Walter Rudin: *Lectures on the edge-of-the-wedge theorem*
7. Yozo Matsushima: *Holomorphic vector fields on compact Kähler manifolds*
8. Peter Hilton: *Lectures in homological algebra*
9. I. N. Herstein: *Notes from a ring theory conference*
10. Branko Grünbaum: *Arrangements and spreads*
11. Irving Glicksberg: *Recent results on function algebras*
12. Barbara L. Osofsky: *Homological dimensions of modules*
13. Michael Rabin: *Automata on infinite objects and Church's problem*
14. Sigurdur Helgason: *Analysis on Lie groups and homogeneous spaces*
15. R. G. Douglas: *Banach algebra techniques in the theory of Toeplitz operators*
16. Joseph L. Taylor: *Measure algebras*
17. Louis Nirenberg: *Lectures on linear partial differential equations*
18. Avner Friedman: *Differential games*
19. Béla Sz.-Nagy: *Unitary dilations of Hilbert space operators and related topics*
20. Hyman Bass: *Introduction to some methods of algebraic K-theory*
21. Wilhelm Ṣtoll: *Holomorphic functions of finite order in several complex variables*
22. O. T. O'Meara: *Lectures on linear groups*
23. Mary Ellen Rudin: *Lectures on set theoretic topology*
24. Melvin Hochster: *Topics in the homological theory of modules over commutative rings*
25. Karl. W. Gruenberg: *Relation modules of finite groups*

1. Introduction

The aim of these lectures is to provide an introduction to recent developments in the theory of class groups and Picard groups. By way of orientation, let us indicate where this theory fits into the general framework of algebra. It is first of all a branch of number theory, since we will be generalizing, to the noncommutative case, the concept of the ideal class group of an algebraic number field. The techniques employed come from three main areas: algebraic number theory, representation theory of algebras and orders, and algebraic K-theory.

As is common with many topics in number theory, part of the appeal of this subject is aesthetic. On the other hand, it has applications to other branches of mathematics. For example, associated with a topological space with fundamental group Π, there is a Swan-Wall invariant which measures whether the space has the same homotopy type as a finite complex. This invariant takes values in the reduced projective class group $\widetilde{K}_0(\mathbf{Z}\Pi)$. There are similar invariants which measure obstructions to the problem of fitting a boundary onto an open manifold. For references, see Milnor [23, page x].

For an algebraic example, consider a finite galois extension E of $\mathbf{Q}$, with group G, and suppose that $E/\mathbf{Q}$ is tamely ramified. Then the ring alg. int. $\{E\}$ is a projective $\mathbf{Z}G$-module, and its class in $\widetilde{K}_0(\mathbf{Z}G)$ measures the obstruction to whether E has a normal integral basis over $\mathbf{Q}$.

As main references for these lectures we cite: Bass [1], CR (Curtis and Reiner [3]), Milnor [23], MO (Reiner [26]).

Throughout, R denotes a Dedekind domain with quotient field F of characteristic 0. For example, if F is an algebraic number field (= finite extension of the rational field $\mathbf{Q}$), then the ring alg. int. $\{F\}$ of all algebraic integers in F is a Dedekind domain with quotient field F. (See CR, §18). Any nonzero prime ideal P of R gives rise to a P-adic valuation of the field F, in which an element $\alpha \in F$ is "small" if P occurs to a high power in the factorization of $R\alpha$ into powers of prime ideals.

By a *prime* of F we mean an equivalence class of valuations of F. The class of the above P-adic valuation will also be denoted by P. However, F may have other valuations as well, not arising from prime ideals of R; in particular, there are the "infinite primes" of F, corresponding to archimedean valuations of F.

For each prime P of F, infinite or not, we may use the P-adic valuation to make F into a metric space, and then we may form its P-adic completion F_P relative to this metric. For P an infinite prime, the completion F_P is either the real field $\mathbf{R}$ or the complex field $\mathbf{C}$, and the prime P is accordingly called *real* or *complex*. On the other hand, for a nonarchimedean or *finite* prime P of an algebraic number field F, the P-adic field F_P is a finite extension of the field $\mathbf{Q}_p$, where p is the unique rational prime in P.

If M is any R-module, we may form its P-adic completion $M_P = R_P \otimes_R M$. Call M an *R-lattice* if M is finitely generated and torsionfree as R-module; in this case, M_P is an R_P-lattice for each P. On the other hand, if M is a finitely generated P-torsion R-module, then $M \cong M_P$. In general, properties of modules can be deduced from those of their completions, as we shall see later.

We recall the definition of the *ideal class group* Cl R of the Dedekind domain R. By an *R-ideal* in F (or *fractional ideal*) we mean an R-lattice $X \subset F$ such that $FX = F$. Two such ideals X, X' are in the same *ideal class* if $X \cong X'$ as R-modules, or equivalently, if $X' = Xa$ for some $a \in F^{\cdot}$ (where $F^{\cdot} = F - \{0\}$). Let $[X]$ denote the ideal class of X; the ideal classes form a group Cl R, relative to the multiplication given by $[X]\,[X'] = [XX']$, where XX' is the set of all finite sums $\Sigma x_i x_i'$, $x_i \in X$, $x_i' \in X'$. When F is an algebraic number field, the group Cl R is finite; this is a special case of the Jordan-Zassenhaus Theorem (see (1.7) below).

Our main purpose here is to study various generalizations of the ideal class group Cl R. In the noncommutative case, it turns out that there are *two* natural generalizations. The first of these considers certain types of *left* ideals, and yields the theory of class groups of orders. The second deals with certain types of *two-sided* ideals, and gives rise to Picard groups. For the commutative case, these two theories will coincide (but it requires a proof that they do so!). Most of these lectures will be devoted to class groups, saving Picard groups for the end.

Now let A be a finite dimensional semisimple F-algebra. We are especially interested in the case where $A = FG$, the group algebra of a finite group G over the field F. By Maschke's Theorem (CR, §10) this group algebra is necessarily semisimple. Left FG-modules are just vector spaces over F, on which the elements of G act from the left as F-linear operators.

Returning to the general case, by an *R-order* in A we mean a subring Λ of A such that

(i) R is contained in the center of Λ,

(ii) Λ is finitely generated as R-module, and

(iii) $F\Lambda = A$, that is, Λ contains an F-basis of A.

(1.1) *Examples.* (1) The *integral group ring* RG is an R-order in the algebra FG.

(2) The ring alg. int. $\{F\}$ is a $\mathbf{Z}$-order in the algebraic number field F.

(3) For any algebraic integer α, the ring $\mathbf{Z}[\alpha]$ is a $\mathbf{Z}$-order in the field $\mathbf{Q}(\alpha)$.

(4) Let $M_n(F)$ denote the ring of all $n \times n$ matrices over F. Then the ring $M_n(R)$ is an R-order in $M_n(F)$.

Hereafter, let Λ be an R-order in the semisimple F-algebra A, where char $F = 0$. We wish to define the class group Cl Λ. By a left *Λ-lattice* M we mean a left Λ-module which is an R-lattice. Each such M is embedded in the A-module $F \otimes_R M$ via $m \longrightarrow 1 \otimes m$, $m \in M$. We always identify M with its image $1 \otimes M$ in $F \otimes_R M$; then $F \otimes_R M$ may be written as FM, the set of F-linear combinations of the elements of M.

A left Λ-*ideal* in A means a left Λ-lattice $M \subset A$ such that $FM = A$. Two such ideals M, N are in the same *ideal class* if $M \cong N$ as Λ-modules, or equivalently, if $M = Nx$ for some $x \in u(A)$. Unfortunately, these ideal classes do not form a group relative to the product $[M]\,[N] = [MN]$. To overcome this difficulty, we must restrict the type of ideals considered.

Two left Λ-modules M, N are in the same *genus* (notation: $M \vee N$) if $M_P \cong N_P$ as left Λ_P-modules for each prime ideal P of R. A left Λ-ideal M in A is *locally free* if it is in the same genus as Λ, that is, $M_P \cong \Lambda_P$ for each P. Let us recall some results on genera (for proofs, see MO, §27).

(1.2) ROITER'S LEMMA. *Let M, N be left Λ-lattices. Then $M \vee N$ if and only if for each finite set S of prime ideals of R, there exists a short exact sequence of Λ-modules $0 \longrightarrow M \longrightarrow N \longrightarrow T \longrightarrow 0$ such that $T_P = 0$ for each $P \in S$.*

(1.3) PROPOSITION. *Let L, M, N be left Λ-lattices such that $L \vee (M \oplus N)$. Then there exist Λ-lattices X, Y such that*

$$L \cong X \oplus Y, \qquad X \vee M, \qquad Y \vee N.$$

(1.4) PROPOSITION. *Let L, M, N be Λ-lattices in the same genus. Then*

$$M \oplus N \cong L \oplus L'$$

for some $L' \vee L$.

Now let M and M' be a pair of locally free Λ-ideals in A; then M and M' are in the same genus as Λ, so by (1.4) we may write

$$M \dotplus M' = \Lambda \dotplus M'' \tag{1.5}$$

for some locally free ideal M''. (We have used the symbol $\dotplus$ to denote external direct sum, in order to avoid any confusion with the concept of internal direct sum of a pair of ideals in a ring.)

Let $\langle M \rangle$ temporarily denote the class of the locally free ideal M. We wish to make these classes into an additive group, by defining

$$\langle M \rangle + \langle M' \rangle = \langle M'' \rangle$$

whenever (1.5) holds. It is clear that $\langle \Lambda \rangle$ acts as zero element. To show the existence of inverses, we note that Λ, Λ and M are in the same genus, so by (1.4) we have $\Lambda \dotplus \Lambda \cong M \dotplus N$ for some locally free ideal N; thus $\langle M \rangle + \langle N \rangle = \langle \Lambda \rangle$, so $\langle N \rangle$ is an additive inverse for $\langle M \rangle$.

There is, however, one difficulty which we have ignored: in order for addition of classes to be well defined, we must know that the isomorphism classes of M and M' uniquely determine the class of M'' in (1.5). This is unfortunately *not* true in general. Swan [39] gave an example in which $\Lambda \dotplus \Lambda \cong \Lambda \dotplus M''$, with M'' not isomorphic to Λ. In this example, Λ is the integral group ring $\mathbf{Z}G$, where G is the generalized quaternion group of order 32.

We shall therefore introduce a new equivalence relation which is slightly weaker than isomorphism. Two left Λ-modules M, N are called *stably isomorphic* if

$$M \dotplus \Lambda^{(k)} \cong N \dotplus \Lambda^{(k)} \quad \text{for some } k,$$

where $\Lambda^{(k)}$ is the external direct sum of k copies of Λ. Let $[M]$ denote the stable isomorphism class of M. We have thus shown

(1.6) THEOREM. *The stable isomorphism classes of locally free ideals form an abelian additive group* Cl Λ, *called the* locally free class group *of* Λ. *Addition is given by*

$$[M] + [M'] = [M''] \quad \text{whenever } M \dotplus M' \cong \Lambda \dotplus M''.$$

The zero element is the class $[\Lambda]$.

For brevity we shall call Cl Λ the *class group* of the R-order Λ. If F is an algebraic number field, the group Cl Λ is finite. This is a consequence of the following basic result (see MO, §26):

(1.7) THEOREM. (JORDAN-ZASSENHAUS). *Let Λ be an R-order in the semisimple F-algebra A, where F is an algebraic number field, and let L be any left Λ-lattice. Then there are only finitely many isomorphism classes of left Λ-lattices M such that $FM \cong FL$ as left A-modules. In particular, there are only a finite number of isomorphism classes of left Λ-ideals in A.*

We shall see later that in the special case where $\Lambda = R$, the locally free class group Cl Λ is isomorphic to the usual ideal class group Cl R. Two other special cases should be mentioned. First, suppose that Λ is a maximal R-order in A (that is, an order not contained in any larger R-order in A). Then (see MO, (11.6)) *every* left Λ-ideal in A is locally free, so Cl Λ is the group of stable isomorphism classes of *all* left Λ-ideals in A.

Second, and more important, suppose that Λ is the integral group ring RG of a finite group G. One is interested in the *projective ideals* of Λ, that is, those left ideals of Λ which are projective as Λ-modules. A basic result states (see CR, §78):

(1.8) THEOREM (SWAN). *Suppose that no rational prime divisor of $|G|$ is a unit in R. Then every projective ideal of RG is locally free.*

As we shall see later, this implies that Cl $RG \cong \widetilde{K}_0(RG)$ if no prime divisor of $|G|$ is a unit in R. In particular, Cl $\mathbf{Z}G \cong \widetilde{K}_0(\mathbf{Z}G)$ for every finite group G.

A left Λ-lattice X is called *locally free of rank n* if $X \vee \Lambda^{(n)}$. We claim that X is necessarily Λ-projective.[1] For consider a Λ-exact sequence

$$(1.9) \qquad 0 \longrightarrow U \longrightarrow V \longrightarrow X \longrightarrow 0.$$

For each prime ideal P of R, there is then a Λ_P-exact sequence

$$(1.10) \qquad 0 \longrightarrow U_P \longrightarrow V_P \longrightarrow X_P \longrightarrow 0.$$

But X_P is free as Λ_P-module, so (1.10) splits. This holds for each P, whence also (1.9) splits (see MO, (3.20) and §5). Therefore X is Λ-projective, as claimed.

The study of locally free modules reduces readily to the case of locally free ideals, by virtue of

---

[1] This also follows from (1.4): since $X \vee \Lambda^{(n)}$, then $\Lambda^{(n)} \dotplus \Lambda^{(n)} \cong X \dotplus Y$ for some $Y \vee \Lambda^{(n)}$.

(1.11) PROPOSITION. *If* $X \vee \Lambda^{(n)}$, *then* $X \cong \Lambda^{(n-1)} \dot{+} M$ *for some locally free left* Λ-*ideal* M *in* A.

PROOF. This follows easily by induction on n, by repeated use of (1.3) and (1.4).

To conclude this section, let us consider the relation between stable isomorphism and isomorphism of locally free Λ-modules, restricting our attention to the case where F is an algebraic number field. It can be shown that two locally free left Λ-modules M, N are stably isomorphic if and only if $M \dot{+} \Lambda \cong N \dot{+} \Lambda$. This follows from Jacobinski [15], [16] (see also Fröhlich [8]), by using a theorem of Eichler. It also follows from the Serre-Bass Cancellation Theorem (see Bass [1, Chapter IV, §3] or Swan [40a, Chapter 12]).

As a matter of fact, in most cases which arise in practice, stable isomorphism implies isomorphism for locally free Λ-modules. A *sufficient* condition for this is that A satisfy the Eichler condition relative to R (notation: A = Eichler/R; see MO, (38.1) for the definition).

(1.12) REMARKS. (i) If A is commutative, or if A is a direct sum of matrix algebras over fields, then A = Eichler/R.

(ii) If A is a simple algebra whose center F has at least one complex prime, then A = Eichler/R.

For the case of group algebras, the following result is useful (see MO, p. 344):

(1.13) PROPOSITION. *Let* $A = FG$, *where* G *is a finite group and* F *is an algebraic number field. Then* A = Eichler/R *if* G *has no homomorphic image of any of the following types*:

generalized quaternion group of order $4n$, $n \geq 2$;
binary tetrahedral group of order 24;
binary octahedral group of order 48;
binary icosahedral group of order 120.

2. Explicit Formulas

We shall use the Localization Sequence of algebraic K-theory to obtain explicit formulas for the class group of an order. This approach, due to Wilson [44], avoids the use of a deep theorem of Eichler. Earlier formulas by Jacobinski [16] and Fröhlich [8] were based on Eichler's Theorem. Let us review some basic facts from K-theory. As general references, we cite Bass [1], Milnor [23], and an excellent expository article by Lam-Siu [20].

Let Λ denote an arbitrary ring (not necessarily an order), and let C be some category of finitely generated left Λ-modules. We wish to define the *Grothendieck group* $K_0(C)$ of the category C. To begin with, let $\mathbf{F}$ denote the free abelian group generated by symbols (M), one for each isomorphism class of objects $M \in C$. Let $\mathbf{F}_0$ be the subgroup of $\mathbf{F}$ generated by all expressions $(M) - (M') - (M'')$, one such for each short exact sequence,

$$(2.1) \qquad 0 \longrightarrow M' \longrightarrow M \longrightarrow M'' \longrightarrow 0$$

which belong to the category C. We then define $K_0(C) = \mathbf{F}/\mathbf{F}_0$ and denote by $[M]$ the image of (M) in $K_0(C)$.

Of especial interest is the case where we start with the category $P(\Lambda)$ of all finitely generated projective left Λ-modules. The corresponding Grothendieck group $K_0(P(\Lambda))$ is called the *projective class group* of Λ, and will be denoted simply as $K_0(\Lambda)$. Since every exact sequence (2.1) of objects in $P(\Lambda)$ must split, we sometimes say that $K_0(\Lambda)$ is generated by symbols $[M]$, with $M \in P(\Lambda)$, and relations $[M] = [M'] + [M'']$ whenever $M \cong M' \oplus M''$. It is easily verified that for $M, N \in P(\Lambda)$, we have $[M] = [N]$ in $K_0(\Lambda)$ if and only if M is stably isomorphic to N.

If the Krull-Schmidt Theorem holds for the ring Λ (that is, every finitely generated Λ-module is expressible as a finite direct sum of indecomposable modules, with the summands uniquely determined up to isomorphism and order of occurrence), then stable isomorphism implies isomorphism. Further, suppose that $\Lambda = \Sigma^{\oplus} P_i$ is a decomposition of Λ into indecomposable left ideals, numbered so that $P_1, \ldots, P_k$ are a full set of nonisomorphic modules among the $\{P_i\}$. Then $[P_1], \ldots, [P_k]$ form a free basis for the free abelian group $K_0(\Lambda)$. We remark that the Krull-Schmidt Theorem holds if Λ is a left artinian ring; it also holds for an order Λ_P over a P-adic ring R_P (see MO, Exercises 6.6, 6.7).

Let $\varphi : \Lambda \longrightarrow \Gamma$ be a ring homomorphism (always assumed such that $\varphi(1) = 1$). Then φ determines an additive homomorphism $\varphi_* : K_0(\Lambda) \longrightarrow K_0(\Gamma)$, given by $\varphi_* = \Gamma \otimes_\Lambda \cdot$. In other words,

$$\varphi_* [M] = [\Gamma \otimes_\Lambda M], \qquad [M] \in K_0(\Lambda).$$

We also need the abelian multiplicative group $K_1(\Lambda)$, the *Whitehead group*[2] of Λ, which is analogous to the group $u(\Lambda)$ of units of Λ. Let $GL(\Lambda)$ be the *general linear group* consisting of all $n \times n$ invertible matrices over Λ, for all n, and let $GL'(\Lambda)$ be its commutator subgroup. We set $K_1(\Lambda) = GL(\Lambda)/GL'(\Lambda)$. It is easily checked that $GL'(\Lambda)$ contains all elementary matrices (those that arise from the identity matrix by inserting one off-diagonal entry). An arbitrary $\xi \in K_1(\Lambda)$ is represented by an invertible $n \times n$ matrix X with entries in Λ. If it happens that by elementary row and column operations X can be put into diagonal form $\operatorname{diag}(\lambda_1, \ldots, \lambda_n)$, then ξ is also represented by $\lambda_1 \cdots \lambda_n$, viewed as a 1×1 matrix. If this procedure can be carried out for each X, then we obtain a surjection $u(\Lambda) \longrightarrow K_1(\Lambda)$. This occurs whenever Λ is a field or skewfield, or a discrete valuation ring. We shall need a more general result of this type (see Bass [1, Chapter V, (9.1)]), namely

(2.2) THEOREM (BASS). *There is a surjection $u(\Lambda) \longrightarrow K_1(\Lambda)$ whenever Λ is a semilocal ring.*

We recall that Λ is *semilocal* if $\Lambda/\operatorname{rad}\Lambda$ is a semisimple artinian ring, where $\operatorname{rad}\Lambda$ denotes the Jacobson radical of Λ. In particular, every left artinian ring is semilocal, as is each finite dimensional algebra over a field. Further, every order Λ over a discrete valuation ring R is semilocal, since $\Lambda/\operatorname{rad}\Lambda$ is a finite dimensional algebra over the residue class field of R (see MO, (6.15)).

We make several remarks about Whitehead groups. First, each ring homomorphism $\varphi: \Lambda \longrightarrow \Gamma$ induces a map $\varphi_*: K_1(\Lambda) \longrightarrow K_1(\Gamma)$. Secondly, if $\Gamma = M_n(\Lambda)$ then there is an isomorphism $K_1(\Gamma) \cong K_1(\Lambda)$, gotten by viewing each element of $GL(\Gamma)$ as an element of $GL(\Lambda)$. Finally, let D be a skewfield, and let $D^\cdot = D - \{0\}$ be its multiplicative group. Set $D^\# = D^\cdot/[D^\cdot, D^\cdot]$, the commutator factor group of $D^\cdot$. Each $X \in GL(D)$ may be diagonalized by elementary row operations. The product of the diagonal elements, viewed as element of $D^\#$, is called the Dieudonné determinant of X. We then obtain an isomorphism $\det: K_1(D) \cong D^\#$.

Let us now turn to the Localization Sequence, assuming hereafter that Λ is an R-order in A. A left Λ-module M has *finite homological dimension* if there exists an exact sequence

$$0 \longrightarrow P_n \longrightarrow P_{n-1} \longrightarrow \cdots \longrightarrow P_1 \longrightarrow P_0 \longrightarrow M \longrightarrow 0 \tag{2.3}$$

in which each P_i is Λ-projective. We set

$$\chi(M) = [P_0] - [P_1] + \cdots + (-1)^n [P_n] \in K_0(\Lambda),$$

assuming that M and each P_i are finitely generated. By repeated use of Schanuel's Lemma, one shows that $\chi(M)$ is independent of the choice of the sequence (2.3).

We denote by $T(\Lambda)$ the category of all finitely generated R-torsion[3] left Λ-modules of finite homological dimension. Then we form the Grothendieck group $K_0(T(\Lambda))$. It turns out that χ induces a homomorphism $K_0(T(\Lambda)) \longrightarrow K_0(\Lambda)$, again denoted by χ. Next, we define $\mu: K_1(A) \longrightarrow K_0(T(\Lambda))$ as follows: given a locally free left Λ-ideal M in A, we may choose a nonzero $r \in R$ such that $rM \subset \Lambda$. Since there is an exact sequence

---

[2] Some authors use this term to denote $K_1(ZG)/\{\pm G\}$ for the case where $\Lambda = ZG$.

[3] This means that for $M \in T(\Lambda)$, there exists a nonzero $a \in R$ such that $aM = 0$.

$$0 \longrightarrow M \xrightarrow{r} \Lambda \longrightarrow \Lambda/rM \longrightarrow 0,$$

it follows that $\Lambda/rM \in T(\Lambda)$. We define

$$[\Lambda//M] = [\Lambda/rM] - [\Lambda/r\Lambda] \in K_0(T(\Lambda)).$$

It is easily shown that $[\Lambda//M]$ is independent of the choice of r. To define $\mu : K_1(A) \longrightarrow K_0(T(\Lambda))$, we note that since A is semilocal, each element of $K_1(A)$ is of the form $\bar{a}$ for some $a \in u(A)$. We set

$$\mu(\bar{a}) = [\Lambda//\Lambda a] \in K_0(T(\Lambda)), \quad a \in u(A).$$

Then μ turns out to be a well-defined homomorphism.

Finally, we observe that the inclusion $\Lambda \subset A$ induces homomorphisms

$$\varphi_0 : K_0(\Lambda) \longrightarrow K_0(A), \quad \varphi_1 : K_1(\Lambda) \longrightarrow K_1(A).$$

We have now the following basic result (Bass [1, Chapter IX, 6.3]):

(2.4) THEOREM (LOCALIZATION SEQUENCE). *Let Λ be an R-order in the F-algebra A. Then there is an exact sequence of abelian groups*

$$K_1(\Lambda) \xrightarrow{\varphi_1} K_1(A) \xrightarrow{\mu} K_0(T(\Lambda)) \xrightarrow{\chi} K_0(\Lambda) \xrightarrow{\varphi_0} K_0(A).$$

It should be stressed that the above result is a rather formal one, in that it does not rely on deep properties of orders or algebraic number fields, but depends instead on relatively straightforward (albeit detailed) arguments based on the definitions of the groups K_0 and K_1, and some of their properties. Let us make one comment about the mysterious term $K_0(T(\Lambda))$. For each $M \in T(\Lambda)$, we may decompose the R-torsion module M into a direct sum of its P-torsion submodules, with P ranging over all prime ideals of R. However, the P-torsion submodule of M is isomorphic to M_P, and $M_P = 0$ a.e.,[4] so we have $M = \sum^{\oplus} M_P$. Further, $M_P \in T(\Lambda_P)$ since $M \in T(\Lambda)$. This yields an isomorphism

$$K_0(T(\Lambda)) \cong \sum_P^{\oplus} K_0(T(\Lambda_P)), \tag{2.5}$$

which will be extremely useful for us.

Returning to the sequence in (2.4), we set $\widetilde{K}_0(\Lambda) = \ker \varphi_0$, the *reduced projective class group*. Thus the sequence

$$1 \longrightarrow K_1(A)/\operatorname{im} K_1(\Lambda) \xrightarrow{\mu} K_0(T(\Lambda)) \xrightarrow{\chi} \widetilde{K}_0(\Lambda) \longrightarrow 0 \tag{2.6}$$

is exact, with μ, χ induced from those in (2.4). Likewise, there is an exact sequence

$$1 \longrightarrow \sum^{\oplus} K_1(A_P)/\operatorname{im} K_1(\Lambda_P) \longrightarrow \sum^{\oplus} K_0(T(\Lambda_P)) \longrightarrow \sum^{\oplus} \widetilde{K}_0(\Lambda_P) \longrightarrow 0, \tag{2.7}$$

where in the direct sums P ranges over all nonzero prime ideals of R. By (2.5), the middle terms in (2.6) and (2.7) are isomorphic. We shall construct homomorphisms λ_0, λ' making the following diagram commute:

$$\begin{array}{ccccccccc} 1 & \longrightarrow & K_1(A)/\operatorname{im} K_1(\Lambda) & \longrightarrow & K_0(T(\Lambda)) & \longrightarrow & \widetilde{K}_0(\Lambda) & \longrightarrow & 0 \\ & & \downarrow{\scriptstyle \lambda'} & & \downarrow{\scriptstyle \text{iso.}} & & \downarrow{\scriptstyle \lambda_0} & & \\ 1 & \longrightarrow & \sum^{\oplus} K_1(A_P)/\operatorname{im} K_1(\Lambda_P) & \longrightarrow & \sum^{\oplus} K_0(T(\Lambda_P)) & \longrightarrow & \sum^{\oplus} \widetilde{K}_0(\Lambda_P) & \longrightarrow & 0. \end{array} \tag{2.8}$$

[4] "a. e." means "almost everywhere", that is, for all but a finite number of prime ideals P.

To begin with, it is clear that the homomorphism $K_0(\Lambda) \longrightarrow K_0(\Lambda_P)$ induces a map $\psi_P : \widetilde{K}_0(\Lambda) \longrightarrow \widetilde{K}_0(\Lambda_P)$. Now let $\xi = [X] - [Y] \in \widetilde{K}_0(\Lambda)$, where $X, Y \in P(\Lambda)$; then $FX \cong FY$, so replacing X by an isomorphic copy, we may assume that in fact $FX = FY$. Then $X_P = Y_P$ a.e. (MO, Exercise 4.6), whence $\psi_P(\xi) = 0$ a.e. Hence the map $\lambda_0 = \Sigma\ \psi_P$ is a well-defined homomorphism from $\widetilde{K}_0(\Lambda)$ into $\Sigma^{\oplus} K_0(\Lambda_P)$. Since the rows of (2.8) are exact, there is then a unique homomorphism λ' making the left hand square commute.

Let us show at once that there is an isomorphism

$$\beta : \mathrm{Cl}\,\Lambda \cong \ker \lambda_0, \tag{2.9}$$

given by $\beta[M] = [\Lambda] - [M]$, $[M] \in \mathrm{Cl}\,\Lambda$. It is clear that β is well defined and monic, and that $\mathrm{im}\,\beta \subset \ker \lambda_0$. To prove that β is surjective, let $\xi \in \ker \lambda_0$. We may write $\xi = [\Lambda^{(k)}] - [X]$ for some k and some $X \in P(\Lambda)$. Then $\psi_P(\xi) = 0$ for each P, whence $X_P \cong \Lambda_P^{(k)}$. Therefore X is locally free of rank k, so by (1.11) we have $X \cong \Lambda^{(k-1)} \dotplus M$ for some locally free ideal M. But then $\xi = [\Lambda] - [M] \in \mathrm{im}\,\beta$. We have now proved that β is a surjection, and it remains to prove that β is a homomorphism. Let M, M', M'' be locally free ideals such that $M \dotplus M' \cong \Lambda \dotplus M''$, so $[M] + [M'] = [M'']$ in $\mathrm{Cl}\,\Lambda$. Then

$$\begin{aligned}\beta[M] + \beta[M'] &= [\Lambda] - [M] + [\Lambda] - [M'] = [\Lambda \dotplus \Lambda] - [M \dotplus M'] \\ &= [\Lambda] - [M''] = \beta[M''],\end{aligned}$$

as desired. This establishes (2.9).

We are going to apply the "Snake Lemma" to (2.8), so let us recall it.

(2.10) SNAKE LEMMA. *Given a commutative diagram of abelian groups*

$$\begin{array}{ccccccc} & & L' & \xrightarrow{\alpha} & L & \xrightarrow{\beta} & L'' & \longrightarrow 0 \\ & & \downarrow{\scriptstyle\theta'} & & \downarrow{\scriptstyle\theta} & & \downarrow{\scriptstyle\theta''} & \\ 0 & \longrightarrow & M' & \xrightarrow[\gamma]{} & M & \xrightarrow[\delta]{} & M'' & \end{array}$$

there is an exact sequence

$$\ker \theta' \xrightarrow{\alpha_*} \ker \theta \longrightarrow \ker \theta'' \xrightarrow{\partial} \mathrm{cok}\,\theta' \longrightarrow \mathrm{cok}\,\theta \xrightarrow{\delta_*} \mathrm{cok}\,\theta''.$$

If α is monic, so is α_*; if δ is epic, so is δ_*.

PROOF. Standard diagram chase. The map ∂ is given by $\gamma^{-1}\theta\beta^{-1}$, suitably interpreted.

Applying this to (2.8), we conclude that $\ker \lambda_0 \cong \mathrm{cok}\,\lambda'$, and λ_0 is surjective. Hence we obtain exact sequences

$$K_1(\Lambda) \longrightarrow K_1(A) \xrightarrow{\lambda'} \textstyle\sum^{\oplus} K_1(A_P)/\mathrm{im}\,K_1(\Lambda_P) \xrightarrow{\partial} \widetilde{K}_0(\Lambda) \xrightarrow{\lambda_0} \sum^{\oplus} \widetilde{K}_0(\Lambda_P) \longrightarrow 0$$

(2.11)

$$\begin{array}{ccccc} & \searrow{\scriptstyle\partial'} & & \nearrow{\scriptstyle\beta} & \\ & & \mathrm{Cl}\,\Lambda & & \\ & \nearrow & & \searrow & \\ 0 & & & & 0 \end{array}$$

with $\mathrm{im}\,\beta = \ker \lambda_0$ and $\beta\partial' = \partial$. Let us describe the map ∂' explicitly. By (2.2), each element in $K_1(A_P)$ is the image of a unit in A_P. Thus each x in the domain of ∂' is expressible as a product $x = \Pi \bar{a}_P$ with $a_P \in u(A_P)$ for all P, and $a_P = 1$ a.e. The image of x in

$\Sigma^{\oplus} K_0(T(\Lambda_P))$ is then $\Sigma[\Lambda_P // \Lambda_P a_P]$, and we must find $x' \in K_0(T(\Lambda))$ which corresponds to this sum under the isomorphism (2.5). We set

$$X = A \cap \left\{ \bigcap_P \Lambda_P a_P \right\}. \tag{2.12}$$

By MO, (5.3), X is a locally free Λ-ideal in A such that $X_P = \Lambda_P a_P$ for all P. The desired x' is therefore $[\Lambda // X]$. Since the image of $[\Lambda // X]$ in $\widetilde{K}_0(\Lambda)$ is $[\Lambda] - [X]$, it follows that $\partial'(x) = [X]$ by virtue of the definition of β.

We shall find it convenient to describe the cokernel of λ' in terms of ideles. Recall that the *idele group* of the field F relative to R is defined by

$$J(F) = \left\{ (\alpha_P) \in \prod u(F_P) : \alpha_P \in u(R_P) \text{ a.e.} \right\}, \tag{2.13}$$

where P ranges over all maximal ideals of R. Analogously, we define the idele group of $K_1(A)$ relative to R by

$$JK(A) = \left\{ (x_P) \in \prod K_1(A_P) : x_P \in \operatorname{im} K_1(\Lambda_P) \text{ a.e.} \right\}. \tag{2.14}$$

This group is independent of the choice of the R-order Λ in A, since if Λ' is another order, then $\Lambda_P = \Lambda'_P$ a.e. We shall also need the *group of unit ideles*

$$UK(\Lambda) = \prod_P K_1(\Lambda_P),$$

which of course depends on Λ. There is an obvious homomorphism $UK(\Lambda) \to JK(A)$, induced by $K_1(\Lambda_P) \to K_1(A_P)$ for each P. We have at once

$$JK(A) = \left\{ \sum_P{}^{\oplus} K_1(A_P) \right\} \cdot \operatorname{im} UK(\Lambda),$$

whence

$$\sum_P{}^{\oplus} \frac{K_1(A_P)}{\operatorname{im} K_1(\Lambda_P)} \cong \frac{JK(A)}{\operatorname{im} UK(\Lambda)}. \tag{2.15}$$

Now we observe that there is a homomorphism $K_1(A) \to \Pi_P K_1(A_P)$, induced by the inclusion $A \subset A_P$ at each P. The image of $K_1(A)$ need not lie in $\Sigma^{\oplus} K_1(A_P)$, as is already clear from the case where $A = \mathbf{Q}$ and $R = \mathbf{Z}$. On the other hand, for each $X \in GL(A)$ we observe that $X_P \in GL(\Lambda_P)$ a.e. Thus there is a well-defined homomorphism

$$K_1(A) \to \left\{ \sum_P{}^{\oplus} K_1(A_P) \right\} \left\{ \prod_P \operatorname{im} K_1(\Lambda_P) \right\} = JK(A),$$

given by the diagonal map $x \to (x)$, $x \in K_1(A)$. It follows at once from (2.15), and the definition of the map λ' in (2.8), that

$$\operatorname{cok} \lambda' \cong \frac{JK(A)}{\operatorname{im} K_1(A) \cdot \operatorname{im} UK(\Lambda)}.$$

Since $\mathrm{Cl}\,\Lambda \cong \ker \lambda_0 \cong \operatorname{cok} \lambda'$, we obtain

(2.16) THEOREM. *There is an isomorphism*

$$\phi : \frac{JK(A)}{\operatorname{im} K_1(A) \cdot \operatorname{im} UK(\Lambda)} \cong \mathrm{Cl}\,\Lambda,$$

given as follows: let $x = (x_P) \in JK(A)$, where x_P is the image of an element $a_P \in u(A_P)$ for each P, and where $a_P = 1$ a. e. Then $\phi(x) = [X]$, where X is the locally free left Λ-ideal in A given by formula (2.12).

The above result is due to Wilson [44], and as we have seen, is a consequence of the Localization Sequence of K-theory. An analogous formula, proved earlier by Fröhlich [8], expresses Cl Λ as a quotient of the idele group $J(A)$. Fröhlich's formula is the idele-theoretic version of an earlier result of Jacobinski [16]; in both cases, Eichler's Theorem plays a vital role. For another K-theory approach, see Wall [45, IV].

In order to eliminate the occurrences of the Whitehead groups in (2.16), we shall use the reduced norm map nr, whose properties we now review (see MO, §9):

(i) If $A = M_n(F)$, then $nr: u(A) \longrightarrow F^{\bullet}$ is just the determinant map, where $F^{\bullet} = F - \{0\}$.

(ii) If $A = M_n(D)$ where D is a skewfield with center F, we may write $(D:F) = k^2$ for some integer k. Let E be a maximal subfield of D. Then there is an embedding $D \longrightarrow M_k(E)$, which gives rise to an embedding $\tau: A \longrightarrow M_{kn}(E)$. Set $nr\, a = \det \tau(a)$, $a \in u(A)$. It can be shown that $nr\, a \in F^{\bullet}$, so we again obtain a multiplicative map $nr: u(A) \longrightarrow F^{\bullet}$.

Hereafter we use the following notation:

$$
\begin{aligned}
&A = A_1 \oplus \cdots \oplus A_s \quad \text{(simple components)},\\
&C = F_1 \oplus \cdots \oplus F_s = \text{center of } A, \quad F_i = \text{center of } A_i,\\
&\mathfrak{O} = R_1 \oplus \cdots \oplus R_s = \text{integral closure of } R \text{ in } C.
\end{aligned} \tag{2.17}
$$

Then for each prime ideal P of R, C_P is the center of A_P, and $\mathfrak{O}_P$ is the integral closure of R_P in C_P. There is a multiplicative map $nr: u(A) \longrightarrow u(C)$, defined componentwise, which extends to a reduced norm $nr: u(A_P) \longrightarrow u(C_P)$ for each P.

Suppose from now on that F is an algebraic number field. For each i, $1 \leqslant i \leqslant s$, let S_i denote the set of all real primes P of F_i such that $(A_i)_P$ is not a full matrix algebra over the real field $(F_i)_P$ (one says that A_i *ramifies* at each such P). Define

$$F_i^+ = \{\alpha \in F_i^{\bullet} : \alpha_P > 0 \text{ for all } P \in S_i\},$$

and let

$$C^+ = F_1^+ \oplus \cdots \oplus F_s^+. \tag{2.18}$$

We quote without proof some vital facts about reduced norms, keeping the above notation.

(2.19) THEOREM. *The reduced norm map $nr: u(A_P) \longrightarrow u(C_P)$ is surjective, and its kernel is the commutator subgroup of $u(A_P)$. If Λ_P is a maximal R_P-order in A_P, then $nr\, u(\Lambda_P) = u(\mathfrak{O}_P)$.*

This result is not too hard to prove; for the image of nr, see MO, Exercise 14.6 and §33. For the kernel of nr, see Nakayama and Matsushima [23a]. The global analogue of (2.19) is considerably deeper:

(2.20) THEOREM. *Let F be an algebraic number field. Then $nr\, u(A) = C^+$, and the kernel of nr is the commutator subgroup of $u(A)$.*

The first assertion is proved in MO, (33.15). The second is a deep result due to Wang, and will not be used here.

The reduced norm map $nr: u(A) \longrightarrow u(C)$ extends uniquely to a map $GL(A) \longrightarrow u(C)$, and thence to a map $nr: K_1(A) \longrightarrow u(C)$. This last map is consistent with the surjection $u(A) \longrightarrow K_1(A)$. Likewise, for each P we obtain a map $K_1(A_P) \longrightarrow u(C_P)$. This yields a homomorphism $nr: JK(A) \longrightarrow J(C)$, where $J(C) = \Pi\, J(F_i)$, using the notation of (2.13) and (2.17). If $x = (\bar{a}_P) \in JK(A)$, where $a_P \in u(A_P)$ for all P, and $a_P \in u(\Lambda_P)$ a.e., then $nr\, x = (nr\, a_P)$. It follows at once from (2.19) that nr maps $JK(A)$ *onto* $J(C)$. Further, if $nr\, x = 1$, then by (2.19) each a_P lies in the commutator subgroup of $u(A_P)$, whence $x = 1$ in $JK(A)$. We have thus shown that $nr: JK(A) \cong J(C)$. Using this, together with the first part of (2.20), we obtain

(2.21) THEOREM. *For F an algebraic number field,*

$$\mathrm{Cl}\,\Lambda \cong J(C)\big/C^{+}\cdot \Pi\, nr\, u(\Lambda_P),$$

where C^+ is given by (2.18), *and where the product extends over all nonzero prime ideals P of R.*

Let us try out this result for the case where A is a simple algebra with center F, and where Λ is a maximal R-order in A. In this case, we know that for each P, Λ_P is a maximal R_P-order in A_P (MO, (11.6)), so by (2.19) $nr\, u(\Lambda_P) = u(R_P)$. Hence we obtain

$$\mathrm{Cl}\,\Lambda \cong J(F)\big/F^{+}\cdot \Pi\, u(R_P). \tag{2.22}$$

The right hand expression may be interpreted in terms of ideals rather than ideles, as follows: let $\alpha = (\alpha_P) \in J(F)$, where $\alpha_P \in \dot{F}_P$ for each P, and $\alpha_P \in u(R_P)$ a.e. Set

$$R\alpha = F \cap \left\{ \bigcap R_P \alpha_P \right\},$$

an R-ideal in F such that $(R\alpha)_P = R_P\alpha_P$ for each P. If $I(R)$ denotes the multiplicative group of R-ideals in F, then we obtain a surjection $\sigma: J(F) \longrightarrow I(R)$. Clearly $\ker \sigma = \Pi\, u(R_P)$, so $J(F)/\Pi\, u(R_P) \cong I(R)$. Under this isomorphism, F^+ maps onto $P^+(R) = \{R\alpha : \alpha \in F^+\}$, a certain group of principal R-ideals in F. Hence (2.22) gives

$$\mathrm{Cl}\,\Lambda \cong I(R)/P^{+}(R). \tag{2.23}$$

If $P^+(R)$ consists of *all* principal ideals, then $\mathrm{Cl}\,\Lambda \cong \mathrm{Cl}\,R$. In general, this is not the case, but there is a homomorphism of $I(R)/P^+(R)$ onto $\mathrm{Cl}\,R$ with kernel a finite elementary abelian 2-group (see MO, Exercise 35.2). Hence, the problem of calculating the class group of a maximal order is solved (or, at least, is reduced to a standard problem in algebraic number theory). In particular, for the special case where $\Lambda = R$ and $A = F$, it is clear that $P^+(R)$ is the group of all principal ideals. Thus by (2.23), $\mathrm{Cl}\,\Lambda$ coincides with the usual ideal class group $\mathrm{Cl}\,R$.

We may finally observe that if Λ is a maximal R-order in the semisimple F-algebra A, then (keeping the notation in (2.17)) we may write $\Lambda = \Lambda_1 \oplus \cdots \oplus \Lambda_s$, where for each i, Λ_i is a maximal R_i-order in A_i (see MO, (10.5)). Therefore,

$$\mathrm{Cl}\,\Lambda \cong \textstyle\sum^{\oplus} \mathrm{Cl}\,\Lambda_i \cong \prod_i I(R_i)/P^{+}(R_i).$$

3. Change of Orders

Throughout this section, let Λ, Λ', Γ denote R-orders in semisimple F-algebras, where F has characteristic zero but need not be an algebraic number field. Let $\rho : \Lambda \longrightarrow \Gamma$ be a homomorphism of R-orders, that is, a ring homomorphism such that $\rho(r) = r$, $r \in R$. We claim that ρ induces a map $\rho_* : \mathrm{Cl}\,\Lambda \longrightarrow \mathrm{Cl}\,\Gamma$, given by

$$\rho_*[M] = [\Gamma \otimes_\Lambda M], \qquad [M] \in \mathrm{Cl}\,\Lambda,$$

where in computing the tensor product we use ρ to make Γ into a right Λ-module. Indeed, for each prime ideal P of R we have

$$(\Gamma \otimes_\Lambda M)_P \cong \Gamma_P \otimes_{\Lambda_P} M_P \cong \Gamma_P \otimes_{\Lambda_P} \Lambda_P \cong \Gamma_P,$$

since M is a locally free Λ-ideal. Thus $\Gamma \otimes_\Lambda M$ is a locally free Γ-module, and can be identified with a Γ-ideal in $F\Gamma$. It is clear that the stable isomorphism class of M determines that of $\Gamma \otimes_\Lambda M$, so ρ_* is well defined, and is obviously a homomorphism. We call ρ a "change of rings" homomorphism, and write $\rho_* = \Gamma \otimes_\Lambda \cdot$.

Keeping the above notation, let $A = F\Lambda$, $B = F\Gamma$. Then ρ induces a homomorphism $A \longrightarrow B$ of F-algebras, and also induces maps

$$K_1(A) \longrightarrow K_1(B), \qquad JK(A) \longrightarrow JK(B), \qquad UK(\Lambda) \longrightarrow UK(\Gamma).$$

It is easily checked that the isomorphism in (2.16) is consistent with the change of rings maps induced by ρ.

Of special interest is the case where ρ is the inclusion map $\Lambda \subset \Lambda'$, where Λ' is a maximal R-order in A. Since

$$(3.1) \qquad \mathrm{Cl}\,\Lambda \cong \frac{JK(A)}{\mathrm{im}\,K_1(A) \cdot \mathrm{im}\,UK(\Lambda)}, \qquad \mathrm{Cl}\,\Lambda' \cong \frac{JK(A)}{\mathrm{im}\,K_1(A) \cdot \mathrm{im}\,UK(\Lambda')},$$

it is clear that the map $\mathrm{Cl}\,\Lambda \longrightarrow \mathrm{Cl}\,\Lambda'$ is surjective. Let us prove this fact directly, without use of (2.16). If M' is any locally free Λ'-ideal in A, then for each P we may write $M'_P = \Lambda'_P a_P$, with $a_P \in u(A_P)$ for all P, and $a_P = 1$ a.e. Define

$$M = A \cap \Big\{\bigcap_P \Lambda_P a_P\Big\},$$

a left Λ-lattice in A such that $M_P = \Lambda_P a_P$ for each P. Then $\Lambda' M = M'$ in A, since the equality holds at each P. But $\Lambda' \otimes_\Lambda M \cong \Lambda' M$, since M is locally free. It follows that $[M] \in \mathrm{Cl}\,\Lambda$ maps onto $[M'] \in \mathrm{Cl}\,\Lambda'$, as desired.

Hereafter let $D(\Lambda)$ denote the kernel of the surjection $\mathrm{Cl}\,\Lambda \longrightarrow \mathrm{Cl}\,\Lambda'$. Since $\mathrm{Cl}\,\Lambda'$ is known from §2, we hope to calculate $\mathrm{Cl}\,\Lambda$ by concentrating on its subgroup $D(\Lambda)$. By (3.1) we have

$$D(\Lambda) \cong \frac{\operatorname{im} K_1(A) \cdot \operatorname{im} UK(\Lambda')}{\operatorname{im} K_1(A) \cdot \operatorname{im} UK(\Lambda)}. \tag{3.2}$$

When F is an algebraic number field, from (2.21) and (2.19) we obtain a simpler formula:

$$D(\Lambda) \cong \frac{C^+ \cdot \Pi u(\mathfrak{O}_P)}{C^+ \cdot \Pi\, nr\, u(\Lambda_P)}, \tag{3.3}$$

where $\mathfrak{O}_P$ is the integral closure of R_P in the center of A_P, and where C^+ is as in (2.18). This latter formula shows that the subgroup $D(\Lambda)$ of $\mathrm{Cl}\,\Lambda$ does not depend on the choice of the maximal order Λ' containing Λ.

It will be desirable to give another characterization of $D(\Lambda)$, valid whether or not F is an algebraic number field.

(3.4) THEOREM. *Let M be a locally free Λ-ideal in A. Then $[M] \in D(\Lambda)$ if and only if there exists a finitely generated left Λ-module X such that $M \oplus X \cong \Lambda \oplus X$.*

PROOF. Let Λ' be a maximal R-order in A containing Λ, and let N be any finitely generated left Λ'-module. Denote by $t(N)$ the R-torsion submodule of N; then $t(N)$ is the kernel of the mapping $N \longrightarrow F \otimes_R N$, so $N/t(N)$ is a Λ'-lattice.

Now let M be any locally free Λ-ideal in A. Then $\Lambda' \otimes_\Lambda M \cong \Lambda' M$, where $\Lambda' M$ is the image of $\Lambda' \otimes_\Lambda M$ in FM.

Suppose now that $M \oplus X \cong \Lambda \oplus X$ for some X. Then there is a left Λ'-isomorphism

$$\psi : \Lambda' M \oplus \Lambda' \otimes X \cong \Lambda' \oplus \Lambda' \otimes X.$$

where $\otimes$ means $\otimes_\Lambda$. Since ψ preserves the R-torsion module, it induces an isomorphism

$$\Lambda' M \oplus Y \cong \Lambda' \oplus Y, \quad \text{where } Y = (\Lambda' \otimes X)/t(\Lambda' \otimes X).$$

But now Y is a Λ'-lattice, and Λ' is a maximal order, so Y is Λ'-projective (see MO, (21.5)). We may then choose a Λ'-lattice Y' such that $Y \oplus Y' \cong \Lambda'^{(k)}$ for some k, and then

$$\Lambda' M \oplus \Lambda'^{(k)} \cong \Lambda' \oplus \Lambda'^{(k)}.$$

Thus $[\Lambda' M] = 0$ in $\mathrm{Cl}\,\Lambda'$, whence $[M] \in D(\Lambda)$ as claimed.

The converse is somewhat harder to prove. Let $[M] \in D(\Lambda)$. Consider the exact sequence of Λ-bimodules:[5]

$$0 \longrightarrow \Lambda \xrightarrow{i} \Lambda' \longrightarrow U \longrightarrow 0,$$

where i is an embedding, and U is an R-torsion Λ-module. Let S be the finite set of prime ideals P of R for which $\Lambda_P \neq \Lambda'_P$. Now $M \vee \Lambda$, so by Roiter's Lemma (1.2), there is an exact sequence

$$0 \longrightarrow M \longrightarrow \Lambda \xrightarrow{g} T \longrightarrow 0 \tag{3.5}$$

of Λ-modules, such that $T_P = 0$ for each $P \in S$. Hence for each P, either $T_P = 0$ or else $\Lambda_P = \Lambda'_P$.

Next we claim that

[5] A Λ-*bimodule* is a two-sided Λ-module L such that $(\lambda_1 x)\lambda_2 = \lambda_1(x\lambda_2)$ for all $\lambda_1, \lambda_2 \in \Lambda$, $x \in L$.

(3.6) $$\Lambda' \otimes_\Lambda T \cong T$$

as left Λ-modules. Since T is an R-torsion module, it suffices to prove that $\Lambda'_P \otimes_{\Lambda_P} T_P \cong T_P$ for each P. But this is clear, since for each P either $T_P = 0$ or else $\Lambda'_P = \Lambda_P$.

From (3.5) we obtain an exact sequence of left Λ'-modules:

$$\operatorname{Tor}_1^\Lambda(\Lambda', T) \xrightarrow{\theta} \Lambda' \otimes_\Lambda M \to \Lambda' \otimes_\Lambda \Lambda \to \Lambda' \otimes_\Lambda T \to 0.$$

Since T is an R-torsion module, so is $\operatorname{Tor}_1^\Lambda(\Lambda', T)$. Hence $\operatorname{im}\theta = 0$, since we have already seen that $\Lambda' \otimes_\Lambda M$ is R-torsionfree. Therefore the sequence of left Λ-modules

(3.7) $$0 \to \Lambda' \otimes_\Lambda M \to \Lambda' \xrightarrow{g'} T \to 0$$

is exact, where we have used (3.6) to replace $\Lambda' \otimes_\Lambda T$ by T.

We now set

$$K = \{(\lambda, \lambda') \in \Lambda \oplus \Lambda' : g(\lambda) = g'(\lambda')\},$$

the *fibre product* or *pullback* of the pair of maps g, g' occurring in (3.5) and (3.7). Then there is a commutative diagram

$$\begin{array}{ccc} K & \xrightarrow{h} & \Lambda \\ {\scriptstyle h'}\downarrow & & \downarrow{\scriptstyle g} \\ \Lambda' & \xrightarrow[g']{} & T \end{array}$$

in which h, h' are induced by the projection maps of $\Lambda \oplus \Lambda'$ onto their respective components. Since g' is epic, so is h; it is easily checked that $\ker h \cong \ker g'$. In this manner, we obtain a commutative diagram with exact rows and columns:

$$\begin{array}{ccccccccc} & & & & 0 & & 0 & & \\ & & & & \downarrow & & \downarrow & & \\ & & & & M & \xrightarrow{1} & M & & \\ & & & & \downarrow & & \downarrow & & \\ 0 & \to & \Lambda' \otimes_\Lambda M & \to & K & \to & \Lambda & \to & 0 \\ & & \downarrow{\scriptstyle 1} & & \downarrow & & \downarrow & & \\ 0 & \to & \Lambda' \otimes_\Lambda M & \to & \Lambda' & \to & T & \to & 0 \\ & & & & \downarrow & & \downarrow & & \\ & & & & 0 & & 0 & & \end{array}$$

Consider its completion at an arbitrary P. When $T_P = 0$, we see that $K_P \cong \Lambda_P \oplus \Lambda'_P$, and both sequences

(3.8) $$0 \to M_P \to K_P \to \Lambda'_P \to 0, \quad 0 \to (\Lambda' \otimes M)_P \to K_P \to \Lambda_P \to 0$$

split. On the other hand, when $T_P \neq 0$ then $\Lambda_P = \Lambda'_P$, so both Λ'_P and Λ_P are Λ_P-projective, and again both sequences in (3.8) split. It follows at once (see MO, (3.20) and §5) that both of the sequences

$$0 \to M \to K \to \Lambda' \to 0, \quad 0 \to \Lambda' \otimes M \to K \to \Lambda \to 0$$

are split. Therefore

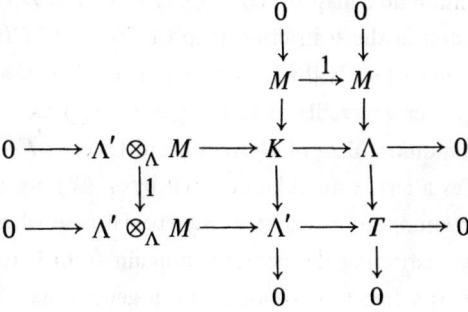

(3.9) $$M \oplus \Lambda' \cong K \cong \Lambda' \otimes_\Lambda M \oplus \Lambda$$

as left Λ-modules. Finally, since $[M] \in D(\Lambda)$ we know that $\Lambda' \otimes_\Lambda M$ is stably isomorphic to Λ', so there exists an integer k such that

$$\Lambda' \otimes M \oplus \Lambda'^{(k)} \cong \Lambda' \oplus \Lambda'^{(k)}.$$

Adding $\Lambda'^{(k)}$ to both sides of (3.9), we obtain

$$M \oplus \Lambda'^{(k+1)} \cong \Lambda \oplus \Lambda'^{(k+1)},$$

which completes the proof of the theorem.

A slightly stronger version of (3.4) is proved in Jacobinski [16]; see also Endo and Miyata [4, I].

From (3.4) we see at once that the subgroup $D(\Lambda)$ of $\mathrm{Cl}\,\Lambda$ does not depend on the choice of Λ'. Even more important, however, is the following consequence (Reiner [27]):

(3.10) COROLLARY. *Let $\rho : \Lambda \longrightarrow \Gamma$ be a homomorphism of R-orders. Then the change of rings map $\rho_* : \mathrm{Cl}\,\Lambda \longrightarrow \mathrm{Cl}\,\Gamma$ carries $D(\Lambda)$ into $D(\Gamma)$.*

PROOF. Let $[M] \in D(\Lambda)$, so $M \oplus X \cong \Lambda \oplus X$ for some Λ-module X. Therefore $\Gamma \otimes M \oplus \Gamma \otimes X \cong \Gamma \oplus \Gamma \otimes X$, where $\otimes$ means $\otimes_\Lambda$. But then by (3.4) we have $[\Gamma \otimes M] \in D(\Gamma)$, that is, $\rho_*[M] \in D(\Gamma)$ as desired.

In particular, let H be a subgroup of the finite group G, and let $\Lambda = RH$, $\Gamma = RG$. The inclusion $\Lambda \subset \Gamma$ gives a change of rings map $\mathrm{Cl}\,RH \longrightarrow \mathrm{Cl}\,RG$, defined by

$$[M] \longrightarrow [RG \otimes_{RH} M], \qquad [M] \in \mathrm{Cl}\,RH.$$

However, $RG \otimes_{RH} M$ is precisely the *induced* RG-module obtained from M, and is usually denoted by M^G. Thus, the induction map $\mathrm{Cl}\,RH \longrightarrow \mathrm{Cl}\,RG$ carries $D(RH)$ into $D(RG)$.

We shall also be interested in the restriction map $\mathrm{Cl}\,RG \longrightarrow \mathrm{Cl}\,RH$, whose existence we now demonstrate. The crucial fact is that RG is free as left RH-module. Let us discuss this situation in somewhat greater generality than the group ring case.

Let $\rho : \Lambda \longrightarrow \Gamma$ be a homomorphism of R-orders, and let $A = F\Lambda$, $B = F\Gamma$. Each left Γ-module X may be viewed as a left Λ-module, denoted by $\rho^*(X)$, by letting Λ act on X via ρ. We call ρ^* the *restriction operator,* since in the special case where ρ is an inclusion, $\rho^*(X)$ is obtained from X by restricting the operator domain from Γ to Λ. Suppose now that $\rho^*(\Gamma) \cong \Lambda^{(n)}$, that is, Γ is a free left Λ-module on n generators. Then $\rho^* : P(\Gamma) \longrightarrow P(\Lambda)$, where as in §2, $P(\Gamma)$ is the category of all finitely generated projective left Γ-modules. Hence ρ^* induces homomorphisms $K_0(\Gamma) \longrightarrow K_0(\Lambda)$, $\widetilde{K}_0(\Gamma) \longrightarrow \widetilde{K}_0(\Lambda)$, and there is a commutative diagram with exact rows:

$$\begin{array}{ccccccccc} 0 & \longrightarrow & \mathrm{Cl}\,\Gamma & \longrightarrow & \widetilde{K}_0(\Gamma) & \xrightarrow{\ 0\ } & \sum^\oplus \widetilde{K}_0(\Gamma_P) & \longrightarrow & 0 \\ & & & & \downarrow{\scriptstyle \rho^*} & & \downarrow{\scriptstyle \rho^*} & & \\ 0 & \longrightarrow & \mathrm{Cl}\,\Lambda & \longrightarrow & \widetilde{K}_0(\Lambda) & \longrightarrow & \sum^\oplus \widetilde{K}_0(\Lambda_P) & \longrightarrow & 0 \end{array}$$

There is then a well-defined homomorphism $\mathrm{res} : \mathrm{Cl}\,\Gamma \longrightarrow \mathrm{Cl}\,\Lambda$ making the left hand square commute. The restriction map *res* may be described explicitly as follows: if M is any locally free Γ-ideal in B, then for each P, $M_P \cong \Gamma_P$; therefore $\rho^*(M_P) \cong \Lambda_P^{(n)}$, which shows that $\rho^*(M)$ is a locally free left Λ-module of rank n. By (1.11) we have $\rho^*(M) \cong \Lambda^{(n-1)} \dotplus N$ for some locally free Λ-ideal N in Λ. Then res $[M] = [N] \in \mathrm{Cl}\,\Lambda$.

We may at once prove Matchett's [22a] result that $\mathrm{res}\, D(\Gamma) \subset D(\Lambda)$. Keeping the preceding notation, let $[M] \in D(\Gamma)$, so there exists a Γ-module X for which $M \oplus X \cong \Gamma \oplus X$. Therefore

$$\rho^*(M) \oplus \rho^*(X) \cong \rho^*(\Gamma) \oplus \rho^*(X),$$

that is,

$$N \oplus \Lambda^{(n-1)} \oplus \rho^*(X) \cong \Lambda \oplus \Lambda^{(n-1)} \oplus \rho^*(X).$$

Hence $[N] \in D(\Lambda)$ by (3.4), and we have established that $\mathrm{res}: D(\Gamma) \longrightarrow D(\Lambda)$.

In particular, let $\Lambda = RH$ and $\Gamma = RG$, with H a subgroup of the finite group G. Set $G = \bigcup_{i=1}^n Hg_i$; then

$$\Gamma = \sum_{i=1}^{n}{}^{\oplus} \Lambda g_i,$$

so Γ is free of rank n as left Λ-module. We summarize our results:

(3.11) THEOREM. *The inclusion* $RH \subset RG$ *yields an induction map* $\mathrm{ind}: \mathrm{Cl}\, RH \longrightarrow \mathrm{Cl}\, RG$ *and a restriction map* $\mathrm{res}: \mathrm{Cl}\, RG \longrightarrow \mathrm{Cl}\, RH$. *These maps satisfy*

$$\mathrm{ind}\, D(RH) \subset D(RG), \qquad \mathrm{res}\, D(RG) \subset D(RH).$$

This theorem permits one to apply the machinery of Frobenius functors to the calculation of $D(RG)$ and $\mathrm{Cl}\, RG$. Applications of this type may be found in [4], [5], [28], [32].

4. Class Groups of p-Groups

Let G be a finite p-group, where p is a rational prime. The aim of this section is to prove that $D(\mathbf{Z}G)$ is also a (finite) p-group. This result, proved first by Fröhlich [6] for the special case where G is abelian, was established in general by Reiner-Ullom [29], [31]. For simplicity we shall consider here only the case where p is odd. Set $\Lambda = \mathbf{Z}G$, $A = \mathbf{Q}G$, and let Λ' be a maximal $\mathbf{Z}$-order in A. For any finite group G, it is easily shown that $|G|\Lambda' \subset \Lambda \subset \Lambda'$ (see MO, (41.1)). Hence in our case it follows that $\Lambda_q = \Lambda'_q$ for each rational prime $q \neq p$.

We shall need some facts about the decomposition of A into its simple components $\{A_i\}$. In terms of the notation in (2.17), each field F_i is a cyclotomic field $\mathbf{Q}(\omega)$ for some p^nth root of unity ω. Further, there is exactly one simple component, say A_1, for which $F_1 = \mathbf{Q}$; and in fact, $A_1 = F_1 = \mathbf{Q}$. Finally, each A_i is a full matrix algebra over F_i. These results are due to Schilling (see MO, (41.9)); for another proof, see Feit [5a, (14.5)].

From algebraic number theory, we cite

(4.1) PROPOSITION. *Let $F_i = \mathbf{Q}(\omega)$, where ω is a primitive p^nth root of 1, with $n \geq 1$. Let $R_i =$ alg. int. $\{F_i\}$. Then there is a unique prime ideal P_i of R_i containing p, and $R_i/P_i \cong \mathbf{Z}/p\mathbf{Z}$. Given any element $\alpha \in R_i$ prime to P_i, there exists a unit $u \in R_i$ such that $\alpha u \equiv 1 \pmod{P_i}$.*

PROOF. The uniqueness of P_i, and the isomorphism $R_i/P_i \cong \mathbf{Z}/p\mathbf{Z}$, both follow from the fact that p is completely ramified in F_i. Now let $\alpha \in R_i$ be prime to P_i; then $\alpha \equiv m \pmod{P_i}$ for some $m \in \mathbf{Z}$ with $(p, m) = 1$. Set $v = (\omega^m - 1)/(\omega - 1)$; then $v \in u(R_i)$ and $v \equiv m \pmod{P_i}$, since $P_i = (1 - \omega)R_i$. We need only choose $u = v^{-1}$, and then $\alpha u \equiv 1 \pmod{P_i}$, as desired.

Since $A_i \cong M_{n_i}(F_i)$ for $1 \leq i \leq s$, no real prime of F_i can ramify in A_i. Hence the group C^+ of (2.18) is precisely $u(C)$. From (3.3) we have

$$D(\Lambda) \cong C^+ \cdot \prod nr\, u(\Lambda'_q) \Big/ C^+ \cdot \prod nr\, u(\Lambda_q)$$
$$\cong \prod u(\mathfrak{O}_q) \Big/ \left\{ \prod nr\, u(\Lambda_q) \right\} \left\{ C^+ \cap \prod u(\mathfrak{O}_q) \right\},$$

where q ranges over all rational primes. However, $C^+ \cap \prod u(\mathfrak{O}_q) = u(\mathfrak{O})$. Further, for $q \neq p$ we have $nr\, u(\Lambda_q) = nr\, u(\Lambda'_q) = u(\mathfrak{O}_q)$. Hence we obtain

$$D(\Lambda) \cong u(\mathfrak{O}_p) \cdot X \Big/ u(\mathfrak{O}) \cdot nr\, u(\Lambda_p) \cdot X,$$

where $X = \prod_{q \neq p} u(\mathfrak{O}_q)$. This implies at once that $D(\Lambda)$ is a homomorphic image of the group

$$H = u(\mathfrak{O}_p)\big/u(\mathfrak{O}) \cdot nr\, u(\Lambda_p).$$

Since $\mathfrak{O} = \Sigma^{\oplus} R_i$, we obtain $\mathfrak{O}_p = \Sigma^{\oplus} (R_i)_p$. By (4.1), the p-adic completion $(R_i)_p$ is the same as the P_i-adic completion $\hat{R}_i$ of R_i; we shall use $\hat{F}_i$ to denote the P_i-adic completion of F_i, for convenience. Thus we have

$$u(\mathfrak{O}_p) = \sum_{i=1}^{s}{}^{\oplus}\, u(\hat{R}_i).$$

Now let $\alpha = \Sigma\, \alpha_i \in u(\mathfrak{O}_p)$, with each $\alpha_i \in u(\hat{R}_i)$. Then $\alpha_i \equiv a_i \pmod{P_i}$ for some $a_i \in R_i$. Choose $t \in Z$ with $(t, p) = 1$ such that $ta_1 \equiv 1 \pmod{P_1}$. Viewing t as an element of $u(\Lambda_p)$, we note that the reduced norm $nr\, t$ has component t in $(A_1)_p$. Let us replace α by $\alpha \cdot nr\, t$, which does not affect the coset $\bar{\alpha} \in H$; this replacement has the effect of making $\alpha_1 \equiv 1 \pmod{P_1}$. On the other hand, for each $i > 1$, we can multiply α_i by a factor from $u(\mathfrak{O})$ without changing the coset $\bar{\alpha}$. By (4.1), we can then make $\alpha_i \equiv 1 \bmod P_i$ for $i > 1$. We have thus shown that each element of H is of the form $\bar{\alpha}$, where $\alpha = \Sigma\, \alpha_i$ and where $\alpha_i \equiv 1 \pmod{P_i}$ for each i.

Let rad Λ_p denote the Jacobson radical of the $\mathbf{Z}_p$-order Λ_p. Then (see MO, (6.15)) for large n we have

$$(\text{rad}\ \Lambda_p)^n \subset p \cdot \Lambda_p \subset \text{rad}\ \Lambda_p, \qquad p^n \Lambda'_p \subset \Lambda_p.$$

It follows at once that if $z \in \Lambda'_p$ is such that $z - 1 \in \text{rad}\ \Lambda'_p$, then $z^{p^n} - 1 \in \text{rad}\ \Lambda_p$ for large n, and therefore $z^{p^n} \in u(\Lambda_p)$. We shall use this fact in a moment.

We may write

$$A_p \cong \sum{}^{\oplus} \{M_{n_i}(F_i)\}_p = \sum{}^{\oplus} M_{n_i}(\hat{F}_i),$$

where $\hat{F}_i$ is the P_i-adic completion of F_i. Up to an inner automorphism of $M_{n_i}(\hat{F}_i)$, each maximal $\hat{R}_i$-order in $M_{n_i}(\hat{F}_i)$ is of the form $M_{n_i}(\hat{R}_i)$ (see MO, (17.4)). Hence we have $\Lambda'_p \cong \Sigma^{\oplus} M_{n_i}(\hat{R}_i)$, from which we deduce at once (see MO, (17.5)) that

$$\text{rad}\ \Lambda'_p \cong \sum{}^{\oplus} P_i \cdot M_{n_i}(\hat{R}_i).$$

Now let $\bar{\alpha} \in H$, where $\alpha = \Sigma\, \alpha_i$ and $\alpha_i \equiv 1 \pmod{P_i}$ for each i. Set

$$x_i = \text{diag}(\alpha_i, 1, \ldots, 1) \in M_{n_i}(\hat{R}_i), \qquad 1 \leqslant i \leqslant s,$$

and let $x = \Sigma x_i$. Then $x - 1 \in \text{rad}\ \Lambda'_p$, and $nr\, x = \alpha$. If $1 + \text{rad}\ \Lambda'_p$ denotes the multiplicative group $\{1 + z : z \in \text{rad}\ \Lambda'_p\}$, then by the above discussion the reduced norm map gives a surjection $\nu : 1 + \text{rad}\ \Lambda'_p \longrightarrow H$. Further, for each $z \in 1 + \text{rad}\ \Lambda'_p$ we proved above that $z^{p^n} \in u(\Lambda_p)$ for large n, and therefore $(nr\, z)^{p^n} \in nr\, u(\Lambda_p)$. This shows that $\nu(z)^{p^n} = 1$ in H, and proves that every element of H has order a power of p. But $D(\mathbf{Z}G)$ is a homomorphic image of H, whence $D(\mathbf{Z}G)$ has this same property. Therefore $D(\mathbf{Z}G)$ is a p-group as claimed, since $D(\mathbf{Z}G)$ is already known to be a finite group.

The proof for the case $p = 2$ is slightly more complicated, and may be found in [29]. We should caution that the obvious generalizations of the above result do *not* hold. Thus, if R is a ring of algebraic integers and G is a p-group, then $D(RG)$ need not be a p-group. On the other hand, if G is an arbitrary finite group, then $|D(\mathbf{Z}G)|$ may have prime divisors other than those occurring in $|G|$ (see Ullom [42]).

5. Mayer-Vietoris Sequences

The explicit formulas for $D(\Lambda)$ in §3 are not well suited for calculations, although they do yield results such as that in §4, and Ullom's [43] estimates for the exponent of $D(\Lambda)$. In practice, most computations are based on Milnor's Mayer-Vietoris sequence, or upon the modification thereof given in Reiner-Ullom [30].

We start with a *fibre product diagram* (or *pullback* diagram)

$$(5.1)\qquad \begin{array}{ccc} \Lambda & \xrightarrow{f_1} & \Lambda_1 \\ {\scriptstyle f_2}\downarrow & & \downarrow{\scriptstyle g_1} \\ \Lambda_2 & \xrightarrow[g_2]{} & \bar{\Lambda} \end{array}$$

of rings and ring homomorphisms. This means that there is a ring isomorphism

$$\Lambda \cong \{(\lambda_1, \lambda_2) : \lambda_i \in \Lambda_i,\ g_1(\lambda_1) = g_2(\lambda_2)\},$$

with f_i the projection map $\Lambda_1 \oplus \Lambda_2 \to \Lambda_i$, $i = 1, 2$. The following basic theorem is due to Milnor [23] (see also Bass [1]):

(5.2) THEOREM. *Given a fibre product diagram* (5.1) *with either* g_1 *or* g_2 *surjective, there is an exact sequence of additive groups*:

$$(5.3)\quad K_1(\Lambda) \to K_1(\Lambda_1) \oplus K_1(\Lambda_2) \to K_1(\bar{\Lambda}) \to K_0(\Lambda) \to K_0(\Lambda_1) \oplus K_0(\Lambda_2) \to K_0(\bar{\Lambda}).$$

(*Furthermore, if both* g_1 *and* g_2 *are surjective, the sequence*

$$(5.4)\qquad K_2(\Lambda) \to K_2(\Lambda_1) \oplus K_2(\Lambda_2) \to K_2(\bar{\Lambda}) \to K_1(\Lambda) \to \cdots$$

is exact.)

The sequence (5.3) is called a Mayer-Vietoris sequence. We shall not make use of (5.4) here, though it does occur in Keating's [18] calculation of $K_1(\mathbf{Z}G)$ for G metacyclic. If Λ is an order, then $K_0(\Lambda)$ is closely related to Cl Λ, and $K_1(\Lambda)$ to the group of units $u(\Lambda)$, so we may expect a sequence of the type (5.3) connecting class groups and unit groups. We shall prove (see Reiner-Ullom [30]):

(5.5) THEOREM. *Let* Λ *be an R-order in a semisimple F-algebra A satisfying the Eichler condition*[6] *relative to R, where F is an algebraic number field, and let* (5.1) *be a*

[6] The purpose of this hypothesis is to guarantee that for locally free Λ-modules, stable isomorphism implies isomorphism.

fibre product diagram of rings in which Λ_1 *and* Λ_2 *are R-orders in semisimple F-algebras. Suppose that* $\bar{\Lambda}$ *is a finite ring, and that either* g_1 *or* g_2 *is surjective. Let us set*

$$u^*(\Lambda_i) = g_i\{u(\Lambda_i)\}, \qquad i = 1, 2.$$

Then there are exact sequences[7]

(5.6) $$1 \longrightarrow u^*(\Lambda_1) \cdot u^*(\Lambda_2) \longrightarrow u(\bar{\Lambda}) \xrightarrow{\delta} \mathrm{Cl}\,\Lambda \xrightarrow{\psi} \mathrm{Cl}\,\Lambda_1 \oplus \mathrm{Cl}\,\Lambda_2 \longrightarrow 0,$$

(5.7) $$1 \longrightarrow u^*(\Lambda_1) \cdot u^*(\Lambda_2) \longrightarrow u(\bar{\Lambda}) \xrightarrow{\delta} D(\Lambda) \xrightarrow{\psi_1} D(\Lambda_1) \oplus D(\Lambda_2) \longrightarrow 0.$$

Here, the map $\mathrm{Cl}\,\Lambda \longrightarrow \mathrm{Cl}\,\Lambda_i$ *is induced by* f_i, $i = 1, 2$, *and* $\delta(u) = [\Lambda u]$, *where*

(5.8) $$\Lambda u = \{(\lambda_1, \lambda_2) : \lambda_i \in \Lambda_i, g_1(\lambda_1) \cdot u = g_2(\lambda_2)\}, \qquad u \in u(\bar{\Lambda}).$$

The hypotheses of (5.5) remain in force for the rest of this section, unless otherwise stated. We set $A_i = F\Lambda_i$, $i = 1, 2$.

We begin the proof with a few simple remarks. For each prime ideal P of R, we may replace each ring in (5.1) by its P-adic completion, thereby obtaining another fibre product diagram. Thus

$$\Lambda_P \cong \{(\lambda_1, \lambda_2) : \lambda_i \in (\Lambda_i)_P, g_1(\lambda_1) = g_2(\lambda_2)\},$$

where now $g_i : (\Lambda_i)_P \longrightarrow \bar{\Lambda}_P$. Likewise, (5.1) yields a fibre product diagram

$$\begin{array}{ccc} A & \longrightarrow & A_1 \\ \downarrow & & \downarrow \\ A_2 & \longrightarrow & 0 \end{array}$$

since $\bar{\Lambda}$ is an R-torsion module (because $\bar{\Lambda}$ is finite). Thus $A \cong A_1 \oplus A_2$; we hereafter identify A with $A_1 \oplus A_2$, which is consistent with identifying Λ with a subring of $\Lambda_1 \oplus \Lambda_2$.

For $i = 1, 2$, let Λ_i' be a maximal R-order in A_i containing Λ. Set $\Lambda' = \Lambda_1' \oplus \Lambda_2'$, a maximal R-order in A containing Λ. If X is a locally free left Λ-lattice of rank n, then $\Lambda_i \otimes_\Lambda X$ is a locally free left Λ_i-lattice of rank n, and may be identified with its image $\Lambda_i X$ computed inside the A-module FX. Likewise, $\bar{\Lambda} \otimes_\Lambda X$ is locally free.

The map $\psi : \mathrm{Cl}\,\Lambda \longrightarrow \mathrm{Cl}\,\Lambda_1 \oplus \mathrm{Cl}\,\Lambda_2$ is defined by $\psi[X] = ([\Lambda_1 X], [\Lambda_2 X])$ for each locally free left Λ-ideal X in A. There is a commutative diagram

$$\begin{array}{ccccccccc} 0 & \longrightarrow & D(\Lambda) & \longrightarrow & \mathrm{Cl}\,\Lambda & \longrightarrow & \mathrm{Cl}\,\Lambda' & \longrightarrow & 0 \\ & & & & \downarrow{\scriptstyle\psi} & & \downarrow{\scriptstyle\text{iso.}} & & \\ 0 & \longrightarrow & D(\Lambda_1) \oplus D(\Lambda_2) & \longrightarrow & \mathrm{Cl}\,\Lambda_1 \oplus \mathrm{Cl}\,\Lambda_2 & \longrightarrow & \mathrm{Cl}\,\Lambda_1' \oplus \mathrm{Cl}\,\Lambda_2' & \longrightarrow & 0, \end{array}$$

so ψ induces a map $\psi_1 : D(\Lambda) \longrightarrow D(\Lambda_1) \oplus D(\Lambda_2)$ making the left-hand square commute. By the Snake Lemma, it follows that $\ker \psi_1 = \ker \psi$, and $\mathrm{cok}\,\psi_1 \cong \mathrm{cok}\,\psi$. Hence if we show that ψ is surjective, we may deduce that ψ_1 is also surjective.

Suppose we are given a locally free Λ_i-ideal X_i in A, $i = 1, 2$. Then for each P we may write

$$(X_i)_P = (\Lambda_i)_P a_P^{(i)}, \quad \text{where } a_P^{(i)} = \text{unit in } (A_i)_P,$$

and $a_P^{(i)} = 1$ a.e., $i = 1, 2$. Let

[7] The expression $u^*(\Lambda_1) \cdot u^*(\Lambda_2)$ stands for $\{u_1 u_2 : u_i \in u^*(\Lambda_i)\}$. It is a subgroup of $u(\bar{\Lambda})$ by virtue of Lemma 5.9 below.

$$X = A \cap \left\{ \bigcap_P \Lambda_P(a_P^{(1)} + a_P^{(2)}) \right\}.$$

Then X is a locally free Λ-ideal in A such that $\Lambda_i X = X_i$, $i = 1, 2$. This proves that ψ is surjective[8], whence so also is ψ_1.

The proof of Theorem 5.5 is based on the following result, which is established in great detail in Reiner-Ullom [30, (4.20)]:

(5.9) LEMMA. *For each* $u \in u(\bar{\Lambda})$, *define*

$$\Lambda u = \{(\lambda_1, \lambda_2) : \lambda_i \in \Lambda_i, g_1(\lambda_1) \cdot u = g_2(\lambda_2)\}.$$

Then

(i) *Each* Λu *is a locally free* Λ-*lattice such that*

$$\Lambda_i \otimes_\Lambda \Lambda u \cong \Lambda_i \cdot \Lambda u = \Lambda_i, \quad i = 1, 2,$$

the last equality holding true inside A.

(ii) *For* $u, u' \in u(\Lambda')$ *we have*

$$\Lambda u \dotplus \Lambda u' \cong \Lambda \dotplus \Lambda uu' \cong \Lambda \dotplus \Lambda u'u.$$

(iii) *Let* $\Lambda' = \Lambda_1' \oplus \Lambda_2'$ *as above. Then*

$$\Lambda' \otimes_\Lambda \Lambda u \cong \Lambda' \cdot \Lambda u = \Lambda',$$

the equality holding true inside A.

(iv) *Let* X *be any projective* Λ-*lattice such that* $\Lambda_i \otimes_\Lambda X = \Lambda_i$, $i = 1, 2$. *Then* $X \cong \Lambda u$ *for some* $u \in u(\bar{\Lambda})$.

(v) *Let* $u^*(\Lambda_i)$ *be defined as in* (5.5), *and let* $u, u' \in u(\bar{\Lambda})$. *Then* $\Lambda u \cong \Lambda u'$ *if and only if* $u' \in u^*(\Lambda_1) \cdot u \cdot u^*(\Lambda_2)$.

Assertion (iii) shows that $[\Lambda u] \in D(\Lambda)$ for each $u \in u(\bar{\Lambda})$, while (ii) gives $[\Lambda u] + [\Lambda u'] = [\Lambda uu']$. Thus the map δ, defined in (5.5) by the formula $\delta(u) = [\Lambda u]$, is a homomorphism from $u(\bar{\Lambda})$ into $D(\Lambda)$. In view of the fact that $\ker \psi_1 = \ker \psi$, we need only establish the exactness of sequence (5.7). By (5.9(i)), we know that $\psi_1 \cdot \delta = 0$. On the other hand, let X be a locally free left Λ-ideal in A such that $\psi_1 [X] = 0$. Then $[\Lambda_i \otimes_\Lambda X] = 0$ in $\mathrm{Cl}\, \Lambda_i$, $i = 1, 2$. Since $A = \mathrm{Eichler}/R$, and $A = A_1 \oplus A_2$, it follows at once that also $A_i = \mathrm{Eichler}/R$, $i = 1, 2$. Therefore $\Lambda_i \otimes_\Lambda X \cong \Lambda_i$, $i = 1, 2$, whence $X \cong \Lambda u$ by (5.9(iv)). This shows that $\ker \psi_1 \subset \operatorname{im} \delta$, and proves that $\ker \psi_1 = \operatorname{im} \delta$.

It remains for us to determine $\ker \delta$. If $u \in u(\bar{\Lambda})$ is such that $\delta(u) = 0$, then $[\Lambda u] = 0$ in $\mathrm{Cl}\, \Lambda$, whence $\Lambda u \cong \Lambda$. Thus by (5.9(v)) we have $u \in u^*(\Lambda_1) \cdot u^*(\Lambda_2)$, as desired. The converse argument also works, and so the exactness of (5.7) is established. It should be emphasized that this proof of Theorem 5.5 is completely routine and formal, *except* for the fact that stable isomorphism implies isomorphism when $A = \mathrm{Eichler}/R$.

In the absence of the Eichler condition, we obtain (see [30, (5.3)]) an exact sequence

(5.10) $\quad 1 \longrightarrow GL_2^*(\Lambda_1) \cdot GL_2^*(\Lambda_2) \longrightarrow GL_2(\bar{\Lambda}) \longrightarrow D(\Lambda) \longrightarrow D(\Lambda_1) \oplus D(\Lambda_2) \longrightarrow 0,$

where $GL_2^*(\Lambda_i)$ is the image of $GL_2(\Lambda_i)$ in $GL_2(\bar{\Lambda})$.

To conclude this section, we indicate how fibre products (5.1) arise in practice. Let I, J be a pair of two-sided ideals in the R-order Λ. The diagram of natural surjections

[8] The fact that ψ is surjective also follows from §3, since there is an inclusion $\Lambda \subset \Lambda_1 \oplus \Lambda_2$ of orders in A.

$$\begin{array}{ccc} \dfrac{\Lambda}{I \cap J} & \longrightarrow & \dfrac{\Lambda}{I} \\ \downarrow & & \downarrow \\ \dfrac{\Lambda}{J} & \longrightarrow & \dfrac{\Lambda}{I+J} \end{array}$$

is always a fibre product diagram. Now FI is a two-sided ideal of A, where $A = F\Lambda$, and $F \otimes_R (\Lambda/I) \cong A/FI$. Hence if Λ/I is R-torsionfree, then Λ/I is an R-order in a semisimple ring summand of A. If also Λ/J is R-torsionfree, and

$$FI \cap FJ = 0, \qquad FI \oplus FJ = A,$$

then the above diagram becomes

$$\begin{array}{ccc} \Lambda & \longrightarrow & \Lambda/I \\ \downarrow & & \downarrow \\ \Lambda/J & \longrightarrow & \Lambda/(I+J). \end{array}$$

Here, the rings Λ, Λ/I and Λ/J are R-orders in semisimple F-algebras, and $\Lambda/(I + J)$ is an R-torsion ring.

6. Calculations

Throughout this section let $\Lambda = \mathbf{Z}G$, $A = \mathbf{Q}G$, where G is a finite group. Let Λ' be a maximal $\mathbf{Z}$-order in A containing Λ, so there is an exact sequence

(6.1) $$0 \longrightarrow D(\Lambda) \longrightarrow \mathrm{Cl}\,\Lambda \longrightarrow \mathrm{Cl}\,\Lambda' \longrightarrow 0.$$

As we showed at the end of §2, the class group $\mathrm{Cl}\,\Lambda'$ is known quite explicitly, and so we shall concentrate here on the calculation of $D(\Lambda)$. It may be mentioned that relatively little is known about determining the additive structure of the group $\mathrm{Cl}\,\Lambda$, once the structures of $D(\Lambda)$ and $\mathrm{Cl}\,\Lambda'$ are known.

Let us begin with the easiest possible example:

(6.2) THEOREM. *Let G be cyclic of prime order p, and let ω be a primitive p-th root of unity over Q. Then $D(\Lambda) = 0$, $\mathrm{Cl}\,\Lambda \cong \mathrm{Cl}\,R$, where $R =$ alg. int. $\{\mathbf{Q}(\omega)\}$.*

PROOF. If g is a generator for G, then $\Lambda = \mathbf{Z} \oplus \mathbf{Z}g \oplus \cdots \oplus \mathbf{Z}g^{p-1}$. We may thus identify Λ with the ring $\mathbf{Z}[x]/(x^p - 1)$ of polynomials in x with integral coefficients, modulo the principal ideal $(x^p - 1)$. We set

$$I = (x-1)\Lambda, \qquad J = \phi(x)\Lambda, \quad \text{where } \phi(x) = \sum_{m=0}^{p-1} x^m.$$

Since $x - 1$ and $\phi(x)$ are relatively prime polynomials, it follows that $I \cap J = 0$. Furthermore, $\phi(x)$ is the minimum polynomial of ω over $\mathbf{Q}$, and $R = \mathbf{Z}[\omega]$ (see CR, §21), so

$$\Lambda/J \cong \mathbf{Z}[x]/(\phi(x)) \cong \mathbf{Z}[\omega] = R.$$

On the other hand

$$\Lambda/I \cong \mathbf{Z}[x]/(x-1) \cong \mathbf{Z}.$$

Finally,

$$\Lambda/(I+J) \cong \mathbf{Z}[x]/(x-1, \phi(x)) \cong \mathbf{Z}/p\mathbf{Z} = \overline{\mathbf{Z}} \quad \text{(say)}.$$

The fibre product diagram (5.11) becomes

$$\begin{array}{ccc} \Lambda & \longrightarrow & \mathbf{Z} \\ \downarrow & & \downarrow \\ R & \underset{g_2}{\longrightarrow} & \overline{\mathbf{Z}}, \end{array}$$

and the map g_2 is given by

$$R = \mathbf{Z}[x]/(\phi(x)) \longrightarrow \mathbf{Z}[x]/(\phi(x), x-1) \cong \mathbf{Z}[\omega]/(\omega - 1) \cong \overline{\mathbf{Z}}.$$

Here, the principal ideal $(\omega - 1)$ of R is precisely the unique prime ideal P of R containing p (see CR, §21).

Since A is commutative, it automatically satisfies the Eichler condition relative tc **Z**. Applying (5.5), we obtain an exact sequence

$$1 \longrightarrow u^*(R) \cdot u^*(\mathbf{Z}) \longrightarrow u(\overline{\mathbf{Z}}) \longrightarrow D(\Lambda) \longrightarrow D(\mathbf{Z}) \oplus D(R) \longrightarrow 0.$$

Now **Z** is a maximal **Z**-order in **Q**, and R is a maximal **Z**-order in $\mathbf{Q}(\omega)$, so we have $D(\mathbf{Z}) = 0$, $D(R) = 0$. Thus

$$D(\Lambda) \cong u(\overline{\mathbf{Z}})/u^*(R) \cdot u^*(\mathbf{Z}),$$

where $u^*(R)$ denotes the image of $u(R)$ in $u(\overline{\mathbf{Z}})$, and likewise for $u^*(\mathbf{Z})$. We need only show that $u^*(R) = u(\overline{\mathbf{Z}})$, and then we will have $D(\Lambda) = 0$, as desired.

Now $u(\overline{\mathbf{Z}}) = \{\overline{m} : m = 1, 2, \ldots, p-1\}$. For each such m, there is a cyclotomic unit $u = (\omega^m - 1)/(\omega - 1) \in u(R)$, such that $u \equiv m \pmod{P}$. Thus $\overline{u} = \overline{m}$ in $u(\overline{\mathbf{Z}})$, which proves that $u(R)$ maps *onto* $u(\overline{\mathbf{Z}})$. Hence $D(\Lambda) = 0$, as claimed. In the same manner, it follows from (5.6) that $\mathrm{Cl}\,\Lambda \cong \mathrm{Cl}\,R$.

This theorem was first proved (in another way) by Rim [33], using results of Reiner. The above proof is due to Milnor. For a second easy example, we show

(6.3) THEOREM. *Let* $V = \langle s \rangle \times \langle t \rangle$ *be a* (2, 2)*-group. Then* $D(\mathbf{Z}V) = 0$.

PROOF. Let

$$\Lambda = \mathbf{Z}V, \qquad I = (t-1)\Lambda, \qquad J = (t+1)\Lambda, \qquad \overline{\mathbf{Z}} = \mathbf{Z}/2\mathbf{Z}.$$

The fibre product diagram (5.11) becomes

$$\begin{array}{ccc} \Lambda & \longrightarrow & \mathbf{Z}[s] \\ \downarrow & & \downarrow \\ \mathbf{Z}[s] & \longrightarrow & \overline{\mathbf{Z}}[s], \end{array}$$

where $\mathbf{Z}[s]$ is the integral group ring of the cyclic group $\langle s \rangle$ of order 2. By (5.7), there is an exact sequence

$$1 \longrightarrow u^*(\mathbf{Z}[s]) \xrightarrow{\gamma} u(\overline{\mathbf{Z}}[s]) \longrightarrow D(\Lambda) \longrightarrow D(\mathbf{Z}[s]) \oplus D(\mathbf{Z}[s]) \longrightarrow 0.$$

But γ is obviously surjective, and $D(\mathbf{Z}[s]) = 0$ by (6.2), whence $D(\Lambda) = 0$ as claimed.

We shall use the preceding result in dealing with the slightly more difficult case of a dihedral group of order 8.

(6.4) THEOREM. [30, p. 329]. *Let*

$$G = \langle x, y : x^4 = 1, y^2 = 1, yxy^{-1} = x^{-1} \rangle,$$

a dihedral group of order 8. *Then* $D(\Lambda) = \mathrm{Cl}\,\Lambda = 0$.

PROOF. Set $R = \mathbf{Z}[i]$, $K = \mathbf{Q}(i)$, where $i^2 = -1$, and let bars denote complex conjugates. Let $H = \langle y \rangle$, a cyclic group of order 2, and let us introduce the *twisted group ring* of H over R:

$$R \circ H = R \oplus Ry, \quad y^2 = 1, \quad y\alpha = \overline{\alpha}y, \quad \alpha \in R.$$

As in (6.3), let $V = \langle s \rangle \times \langle t \rangle$ be a (2, 2)-group. We set $\Lambda = \mathbf{Z}G$, and define the two-sided ideals I, J of Λ by

$$I = (x^2 + 1)\Lambda, \quad J = (x^2 - 1)\Lambda.$$

Since Λ is free as $\mathbf{Z}[x]$-module, it follows readily that $I \cap J = 0$. The ideals I, J are two-sided because $x^2 + 1$ lies in the center of Λ.

The factor ring Λ/J is generated by the images $\bar{x}, \bar{y}$ of the elements x, y, and we have the relations

$$\bar{x}^2 - 1 = 0, \quad \bar{y}^2 = 1, \quad \bar{y}\bar{x}\bar{y}^{-1} = \bar{x}^{-1}.$$

Hence there is an isomorphism $\Lambda/J \cong \mathbf{Z}V$, given by $\bar{x} \longrightarrow s, \bar{y} \longrightarrow t$. On the other hand, in Λ/I we have

$$\bar{x}^2 + 1 = 0, \quad \bar{y}^2 = 1, \quad \bar{y}\bar{x}\bar{y}^{-1} = \bar{x}^{-1}.$$

Therefore there is an isomorphism $\Lambda/I \cong R \circ H$, given by $\bar{x} \longrightarrow i, \bar{y} \longrightarrow y$. Finally,

$$\frac{\Lambda}{(I+J)} = \frac{\mathbf{Z}[x, y]}{(x^2+1, x^2-1)} = \frac{\mathbf{Z}[x, y]}{(x^2-1, 2)} \cong \bar{\mathbf{Z}}V, \quad \text{where } \bar{\mathbf{Z}} = \mathbf{Z}/2\mathbf{Z}.$$

Hence we obtain from (5.11) the fibre product diagram

$$\begin{array}{ccc} \Lambda & \longrightarrow & R \circ H \\ \downarrow & & \downarrow \\ \mathbf{Z}V & \longrightarrow & \bar{\mathbf{Z}}V. \end{array}$$

From this we deduce that $A = \mathbf{Q}\Lambda \cong \mathbf{Q}V \dotplus K \circ H$. Now it is easily seen that there are ring isomorphisms

(6.5) $$\mathbf{Q}V \cong \mathbf{Q} \dotplus \mathbf{Q} \dotplus \mathbf{Q} \dotplus \mathbf{Q}, \qquad K \circ H \cong M_2(\mathbf{Q}) \subset M_2(K),$$

the latter given by

$$\alpha + \beta y \longrightarrow \begin{bmatrix} \alpha & \beta \\ \bar{\beta} & \bar{\alpha} \end{bmatrix}, \qquad \alpha, \beta \in K.$$

It follows that

(6.6) $$nr\,(\alpha + \beta y) = \alpha\bar{\alpha} - \beta\bar{\beta}, \qquad \alpha, \beta \in K.$$

Furthermore,

$$\Lambda' \cong \mathbf{Z} \dotplus \mathbf{Z} \dotplus \mathbf{Z} \dotplus \mathbf{Z} \dotplus M_2(\mathbf{Z}),$$

so $\mathrm{Cl}\,\Lambda' = 0$ and thence $D(\Lambda) = \mathrm{Cl}\,\Lambda$.

By (6.5), A satisfies the Eichler condition relative to $\mathbf{Z}$. Thus there is an exact sequence

$$1 \longrightarrow u^*(\mathbf{Z}V) \cdot u^*(R \circ H) \xrightarrow{\gamma} u(\bar{\mathbf{Z}}V) \longrightarrow D(\Lambda) \longrightarrow D(R \circ H) \oplus D(\mathbf{Z}V) \longrightarrow 0.$$

We have already seen in (6.3) that $D(\mathbf{Z}V) = 0$. We shall prove now that γ is surjective, and that $D(R \circ H) = 0$. This will imply that $D(\Lambda) = 0$, and will complete our proof.

First of all, we claim that $u(\bar{\mathbf{Z}}V)$ has 8 elements, that is, half of the elements of $\bar{\mathbf{Z}}V$ are units. Indeed, since char $\bar{\mathbf{Z}} = 2$ and V is a 2-group, it follows that rad $\bar{\mathbf{Z}}V$ is the augmentation ideal of $\bar{\mathbf{Z}}V$, and that $\bar{\mathbf{Z}}V/\text{rad}\,\bar{\mathbf{Z}}V \cong \bar{\mathbf{Z}}$ (see CR, Exercise 64.1). Since $u(\bar{\mathbf{Z}}V)$ is the inverse image of $u(\bar{\mathbf{Z}})$, it follows that $u(\bar{\mathbf{Z}}V)$ has order 8. However, $u^*(\mathbf{Z}V) = \{1, s, t, st\}$, a subgroup of index 2 in $u(\bar{\mathbf{Z}}V)$. Further, $(1 + i) + y$ is a unit in $R \circ H$ whose image in $u(\bar{\mathbf{Z}}V)$ does not lie in $u^*(\mathbf{Z}V)$. This completes the proof that γ is surjective.

Finally, since $|G| \cdot \Lambda' \subset \Lambda$ (see MO, (41.1)) it follows that for some m we have $2^m \cdot M_2(\mathbf{Z}) \subset R \circ H$, both viewed as subrings of $M_2(\mathbf{Q})$. As in §4, we obtain

$$D(R \circ H) \cong \frac{nr\, GL_2(\mathbf{Z}_2)}{nr\, u(R_2 \circ H) \cdot u(\mathbf{Z})} = \frac{u(\mathbf{Z}_2)}{nr\, u(R_2 \circ H) \cdot u(\mathbf{Z})}.$$

The general element of $u(\mathbf{Z}_2)$ is of the form $1 + 2\theta$ with $\theta \in \mathbf{Z}_2$. However, $(\theta + 1) + \theta y \in u(R_2 \circ H)$, and by (6.6), $nr\,((\theta + 1) + \theta y) = 1 + 2\theta$. Therefore $D(R \circ H) = 0$, and the theorem is proved.

The preceding calculations illustrate the importance of the Mayer-Vietoris sequences in determining class groups. It is clear that the major difficulty in each case is the calculation of the images of $u(\Lambda_1)$ and $u(\Lambda_2)$ in $u(\bar{\Lambda})$, where we have used the notation of (5.5). If the Eichler condition fails to hold, the problem becomes slightly more complicated, but is not significantly harder; see, for example, the proof that $|D(\Lambda)| = |\mathrm{Cl}\,\Lambda| = 2$ when $\Lambda = \mathbf{Z}G$, with G the quaternion group of order 8 (see [30, p. 327]). Finally, we remark that in almost all cases where this fibre product method has been applied, one starts with a group G having a cyclic normal subgroup.

7. Survey of Specific Results

We give here a brief indication, without proofs, of the current state of knowledge about $D(\mathbf{Z}G)$ and Cl $\mathbf{Z}G$ for various groups G. Let us start with the case where G is a p-group. We know from §4 that $D(\mathbf{Z}G)$ is also a p-group, but there is a more precise result due to Ullom [43, (3.1)]:

(7.1) THEOREM. *Let $|G| = p^n$, p an odd prime, $n \geqslant 1$. Then the exponent*[9] *of $D(\mathbf{Z}G)$ divides p^{n-1}. If $|H| = 2^n$ where $n \geqslant 2$, then the exponent of $D(\mathbf{Z}H)$ divides 2^{n-2}.*

Since the cyclic group of order p was handled so easily in (6.2), it seems reasonable to try next the case where G is cyclic of order p^2, or even p^n. Here we already meet major difficulties, arising mainly from the fact that there is just not enough information available about units in cyclotomic fields. The results described below are due to Kervaire and Murthy [19] and Galovich [12]. The *class number* of an algebraic number field F is the order of the ideal class group of alg. int. $\{F\}$. An odd prime p is called *regular* if p does not divide the class number of $Q(\omega)$, where $\omega = \sqrt[p]{1}$. An irregular prime p is said to be *properly irregular* if p does not divide the class number of $Q(\omega + \omega^{-1})$.

(7.2) THEOREM. *Let G_n denote a cyclic group of order p^{n+1}, where p is an odd prime and $n \geqslant 0$. Then*

(i) *If p is regular, then $|D(\mathbf{Z}G_n)| = p^e$, where*

$$e = n^2(p-3)/2 + (n-1)\{(n-1)(p^2-3p+2)/2 + 1\}$$
$$+ \sum_{i=2}^{n-1} (n-i)\{(n-i)p^{i-2}(p-1)^3/2 + p^{i-1}(p-1)/2 + 1\}.$$

Thus in particular, when p is regular we have $|D(\mathbf{Z}G_1)| = p^{(p-3)/2}$.

(ii) *If p is properly irregular, then $|D(\mathbf{Z}G_1)| = p^e$, where $e = (p-3)/2 + \delta_p$, and δ_p is the number of Bernoulli numbers $B_1, B_2, \ldots, B_{(p-3)/2}$ which are divisible by p.*

For the case where p is irregular, it turns out [19] that $|D(\mathbf{Z}G_n)|$ is a multiple of the expression p^e given in (i). However, no explicit formula for $|D(\mathbf{Z}G_n)|$ is known in this case. Some results relating to this problem may also be found in Fröhlich [6, II]; see also Ullom [43b].

Next, we have [30, (8.4)]:

(7.3) THEOREM. *Let G be an abelian p-group of order p^{n+1}, where p is an odd prime. Then $|D(\mathbf{Z}G)|$ is a multiple of p^e, where*

---

9 The *exponent* of a finite group is the least common multiple of the orders of its elements.

$$e = \sum_{r=1}^{n} \{(p^r - 1)/2 - r\}.$$

For further results on the case where G is an abelian p-group, see Fröhlich [6, I, II]. For the case of elementary abelian p-groups, see Wall [46] for $p = 2$, and Galovich [in preparation] for regular odd p.

We have remarked above that for G the quaternion group of order 8, one has $|D(\mathbf{Z}G)| = |\mathrm{Cl}\,\mathbf{Z}G| = 2$ (see [30, p. 327], [22] or [46]). This fact, as well as that in (6.4), has been generalized in an elegant paper of Fröhlich, Keating and Wilson [10], who proved:

(7.4) THEOREM. *Let*

$$G = \langle x, y : x^{2^{n-1}} = 1, y^2 = 1, yxy^{-1} = x^{-1} \rangle$$

be the dihedral group of order 2^n. *Let*

$$H = \langle x, y : x^{2^{n-1}} = y^2, y^4 = 1, yxy^{-1} = x^{-1} \rangle,$$

the generalized quaternion group of order 2^{n+1}. *Then for* $n \geqslant 2$, $|D(\mathbf{Z}G)| = 1$ *and* $|D(\mathbf{Z}H)| = 2$.

Some related results are given in Endo and Miyata [4], [5]. We quote next

(7.5) THEOREM [30, (8.15)]. *Let p, q be distinct primes, and set f = order of p* mod q, $g = (q-1)/f$,

$$N(p, q) = \begin{cases} (p^{f/2} + 1)^g/q, & f \text{ even} \\ (p^f - 1)^{g/2}/q, & f \text{ odd.} \end{cases}$$

(i) *If G is cyclic of order pq, where p and q are odd, then $4\,|D(\mathbf{Z}G)|$ is divisible by* $N(p, q)N(q, p)$.

(ii) *If G is cyclic of order $2p$, where p is odd, then $|D(\mathbf{Z}G)|$ is divisible by* $N(2, p)$.

For explicit results on $D(\mathbf{Z}G)$ in case (ii), see Ullom [42], and also [21], [29], [30]. Cassou-Noguès [2] determined all abelian groups G for which $D(\mathbf{Z}G) = 0$. Generally speaking, $|D(\mathbf{Z}G)|$ is large for G an abelian group of composite order, and we have [30, (8.2)]:

(7.6) THEOREM. *Let $\{G_i\}$ be any sequence of abelian groups of composite order, such that* $|G_i| \longrightarrow \infty$. *Then also* $|D(\mathbf{Z}G_i)| \longrightarrow \infty$.

Turning now to the case of metacyclic groups, we quote a result of Galovich, Reiner and Ullom [13] (see also Lee [21], Pu [24]):

(7.7) THEOREM. *Let*

$$G = \langle x, y : x^p = 1, y^q = 1, yxy^{-1} = x^r \rangle,$$

where p is an odd prime, q any divisor of $p - 1$, and r is a primitive q-th root of 1 mod p. *Let ω be a primitive p-th root of* 1 *over* $\mathbf{Q}$, *and let L be the unique subfield of* $\mathbf{Q}(\omega)$ *such that* $(\mathbf{Q}(\omega) : L) = q$. *Put* $S =$ alg. int. $\{L\}$. *Let* $H = \langle y \rangle$, *a cyclic group of order q. Then there is a surjection*

$$\mathrm{Cl}\,\mathbf{Z}G \longrightarrow \mathrm{Cl}\,S \oplus \mathrm{Cl}\,\mathbf{Z}H$$

whose kernel D_0 *is a cyclic group of order* $q/(q, 2)$. *Further, the sequence*

$$0 \to D_0 \to D(\mathbf{Z}G) \to D(\mathbf{Z}H) \to 0$$

is exact. In particular, $|D(\mathbf{Z}G)| = 1$ *if* G *is a dihedral group of order* $2p$.

This result was extended by Keating [18] to the case where G is metacyclic of order $p^r q$, with p a regular prime. We may also mention the work of Fröhlich [9] and Wilson [44], who determined the 2-primary component of $D(\mathbf{Z}H)$ for the case where H is the generalized quaternion group

$$H = \langle x, y : x^{p^n} = 1, y^4 = 1, yxy^{-1} = x^{-1} \rangle,$$

with p an odd prime; namely, in this case the 2-primary component is an elementary abelian group of rank n.

8. Induction Techniques

The method of induced representations, so important for the theory of group representations, has so far yielded relatively little information about class groups. However, it seems to be the only way of handling groups without cyclic normal subgroups.

Recall that a *hyper-elementary* group is one which is a semidirect product NP of a cyclic normal subgroup N and a subgroup P of prime power order, where $(|N|, |P|) = 1$. The following result is due to Swan [38] (see also [41]):

(8.1) THEOREM. *Let H range over a full set of nonconjugate hyper-elementary subgroups of G. Then the maps*

$$\mathrm{Cl}\,\mathbf{Z}G \longrightarrow \prod_H \mathrm{Cl}\,\mathbf{Z}H, \qquad D(\mathbf{Z}G) \longrightarrow \prod_H D(\mathbf{Z}H),$$

defined by restriction at each H, are monomorphisms.

OUTLINE OF PROOF. Let $G_0(\mathbf{Z}G)$ be the Grothendieck group of the category of all left $\mathbf{Z}G$-lattices, and let ind_H^G denote the induction map. Swan proves that

$$G_0(\mathbf{Z}G) = \sum_H \mathrm{ind}_H^G G_0(\mathbf{Z}H).$$

Further, $G_0(\mathbf{Z}G)$ has a ring structure, arising from "inner" tensor products over $\mathbf{Z}$. Now $\mathrm{Cl}\,\mathbf{Z}G$ may be viewed as a left $G_0(\mathbf{Z}G)$-module, via $[X]\,[M] = [X \otimes_{\mathbf{Z}} M]$ for $[X] \in G_0(\mathbf{Z}G)$, $[M] \in \mathrm{Cl}\,\mathbf{Z}G$. It then follows by elementary reasoning that $\mathrm{Cl}\,\mathbf{Z}G \longrightarrow \Pi\,\mathrm{Cl}\,\mathbf{Z}H$ is monic. The same holds for the D's, since $D(\mathbf{Z}G)$ is also a $G_0(\mathbf{Z}G)$-module by (3.4).

This theorem enables us to get upper bounds on the size of $|\mathrm{Cl}\,\mathbf{Z}G|$ and $|D(\mathbf{Z}G)|$. In particular, let A_n denote the alternating group on n symbols, and S_n the symmetric group. For small n, it turns out that $\mathrm{Cl}\,\mathbf{Z}H = 0$ for each hyper-elementary subgroup H of A_n or S_n. This gives (Reiner-Ullom [32]):

(8.2) THEOREM. $\mathrm{Cl}\,\mathbf{Z}A_n = 0$ *for* $1 \leqslant n \leqslant 5$, *and* $\mathrm{Cl}\,\mathbf{Z}S_n = 0$ *for* $1 \leqslant n \leqslant 4$.

Let us show how to obtain lower bounds on the size of these class groups. We prove (Reiner [28, Theorem 2]):

(8.3) THEOREM. *Let $n \geqslant 5$ and let p be any odd prime*[10] *such that $n/2 < p \leqslant n$. Then*

(i) $|D(\mathbf{Z}S_n)|$ *is a multiple of* $(p-1)/2$.

(ii) $|D(\mathbf{Z}A_n)|$ *is a multiple of* $(p-1)/2$ *if* $p \leqslant n-2$.

---

[10] Such a prime always exists, by Bertrand's Postulate.

(iii) *$|D(\mathbf{Z}A_n)|$ is a multiple of $(p-1)/2\lambda_p$ if $n = p$ or $p+1$, where*

$$\lambda_p = GCD(2, (p-1)/2).$$

Hence

$$D(\mathbf{Z}S_n) \neq 0 \quad \text{for } n \geqslant 5, \qquad D(\mathbf{Z}A_n) \neq 0 \quad \text{for } n \geqslant 7.$$

We begin the proof by establishing

(8.4) LEMMA. *Given finite groups $H_1 \subset G_1 \subset G$ such that G is a semidirect product $N \cdot G_1$, set $H = N \cdot H_1$. Consider the commutative diagram of additive groups, where the arrows denote induction maps:*

$$\begin{array}{ccc} D(\mathbf{Z}H_1) & \xrightarrow{\alpha_*} & D(\mathbf{Z}H) \\ \beta_* \downarrow & & \downarrow \delta_* \\ D(\mathbf{Z}G_1) & \xrightarrow[\gamma_*]{} & D(\mathbf{Z}G). \end{array}$$

Then cok β_* *is a direct summand of* cok δ_*.

PROOF. The inclusion $\gamma : \mathbf{Z}G_1 \longrightarrow \mathbf{Z}G$ gives the induction map γ_* above. But γ is split by the surjection $\varphi : \mathbf{Z}G \longrightarrow \mathbf{Z}G_1$ arising from the homomorphism $G \longrightarrow G/N \cong G_1$, so $\varphi\gamma = 1$. Therefore $\varphi_*\gamma_* = 1$, so $D(\mathbf{Z}G_1)$ is a direct summand of $D(\mathbf{Z}G)$.

Likewise, the inclusion $\alpha : \mathbf{Z}H_1 \longrightarrow \mathbf{Z}H$ is split by $\varphi' : \mathbf{Z}H \longrightarrow \mathbf{Z}H_1$, where φ' is the restriction of φ to $\mathbf{Z}H$. Thus $D(\mathbf{Z}H_1)$ is a direct summand of $D(\mathbf{Z}H)$, and this direct sum decomposition is consistent with that of $D(\mathbf{Z}G)$. Hence cok β_* is a direct summand of cok δ_*, as claimed.

We are now ready to prove (8.3). We shall prove only (i), since analogous arguments work for (ii) and (iii). Let the symmetric group S_n act on the symbols $0, 1, \ldots, n-1$, and let r be a primitive root mod p. We set

(8.5) $$x = (0, 1, \ldots, p-1), \qquad y = (1, r, r^2, \ldots, r^{p-2}) \in S_p,$$

where the entries of the $(p-1)$-cycle y are reduced mod p so as to lie between 1 and $p-1$. Then the normalizer G_1 of the cyclic group $\langle x \rangle$ in S_p is given by

$$G_1 = \langle x, y : x^p = 1, y^{p-1} = 1, yxy^{-1} = x^r \rangle;$$

the normalizer of $\langle x \rangle$ in S_n is $G = G_1 \times S_{n-p}$, where S_{n-p} acts on the symbols $p, p+1, \ldots, n-1$. Since $n/2 < p \leqslant n$, it is clear that $\langle x \rangle$ is a Sylow p-subgroup of S_n, and G is self-normalizing in S_n. Further, for each $a \in S_n - G$ we must have $G^a \cap \langle x \rangle = 1$, since otherwise a normalizes $\langle x \rangle$ and hence lies in G. We intend to apply Lemma 8.4 with

(8.6) $$H_1 = \langle y \rangle \subset G_1 = \langle x, y \rangle \subset G, \qquad N = S_{n-p}, \qquad H = N \times H_1.$$

First, however, let us consider the maps

$$D(\mathbf{Z}H) \xrightarrow{\delta_*} D(\mathbf{Z}G) \xrightarrow{\sigma_*} D(\mathbf{Z}S_n),$$

where δ, σ are inclusion maps. We claim that ker $\sigma_* \subset$ im δ_*. Indeed, let $X \in D(\mathbf{Z}G)$ be such that $\sigma_* X = 0$; then also $\sigma^*(\sigma_* X) = 0$, where $\sigma^* : D(\mathbf{Z}S_n) \longrightarrow D(\mathbf{Z}G)$ is the restriction map. By Mackey's Theorem (see CR, (44.2)), we may write

(8.7) $$\sigma^*(\sigma_* X) = X \oplus \sum_a{}^{\oplus} \{(a \otimes X)_{G \cap G^a}\}^G,$$

where a ranges over a full set of (G, G)-double coset representatives in S_n, excluding 1. Each such a lies in $S_n - G$, so by the remarks in the preceding paragraph, $G \cap G^a$ is a p'-subgroup of G. But G is p-solvable and G is a semidirect product $\langle x \rangle \cdot H$, whence by P. Hall's theorem (see Huppert [14a, p. 660, Theorem 1.7]) it follows that $G \cap G^a$ is G-conjugate to a subgroup of H. However, since $X \in D(\mathbf{Z}G)$ we have $a \otimes X \in D(\mathbf{Z}G^a)$, whence by (3.11) the restriction of $a \otimes X$ to $G \cap G^a$ lies in $D(\mathbf{Z}(G \cap G^a))$. Hence by (3.11) and transitivity of the induction map, we conclude that

$$\{(a \otimes X)_{G \cap G^a}\}^G \in \operatorname{im} \delta_*,$$

for each $a \in S_n - G$. Since the left-hand expression in (8.7) equals zero, it thus follows from (8.7) that $X \in \operatorname{im} \delta_*$. This completes the proof that $\ker \sigma_* \subset \operatorname{im} \delta_*$.

The above implies that $D(\mathbf{Z}G)/\operatorname{im} \delta_*$ is a factor group of $D(\mathbf{Z}G)/\ker \sigma_*$; this latter group is embedded in $D(\mathbf{Z}S_n)$ by means of σ_*. Therefore $|\operatorname{cok} \delta_*|$ divides $|D(\mathbf{Z}S_n)|$. However, using (8.4) and the notation of (8.6), we see that $|\operatorname{cok} \delta_*|$ is a multiple of $|\operatorname{cok} \beta_*|$, where $\beta_* : D(\mathbf{Z}H_1) \longrightarrow D(\mathbf{Z}G_1)$. By (7.7) we know that $|\operatorname{cok} \beta_*|$ is a multiple of $(p-1)/2$, and thus (8.3(i)) is established.

REMARKS. 1. Endo-Miyata [5] proved (i) for the case where $n = p$ or $p + 1$, and also proved the analogue of (iii) for $|\mathrm{Cl}\, \mathbf{Z}A_n|$. It is easily seen that $|\mathrm{Cl}\, \mathbf{Z}A_n|$ is a multiple of $h(p-1)/2\lambda_p$, where h is the class number of the quadratic extension of $\mathbf{Q}$ contained in $\mathbf{Q}(\sqrt[p]{1})$.

2. As remarked in [43], $\mathrm{Cl}\, \mathbf{Z}S_n = D(\mathbf{Z}S_n)$ for all n, since $\mathbf{Q}S_n$ is a direct sum of full matrix algebras over $\mathbf{Q}$.

3. The results in (8.3) have been obtained in another way by Ullom [43a], who showed in fact that an odd prime p divides $|\mathrm{Cl}\, \mathbf{Z}S_n|$ if and only if $p \leqslant n/2$. He also proved that $|\mathrm{Cl}\, \mathbf{Z}A_6| = 3$, and that $|\mathrm{Cl}\, \mathbf{Z}A_n|$ is divisible by every odd prime $p \leqslant n/2$. Ullom's proofs are based on the study of a certain subgroup $T(G)$ of $D(\mathbf{Z}G)$, defined as follows: for an arbitrary finite group G, set $\Lambda = \mathbf{Z}G$, $n = |G|$, $\sigma_G = \Sigma_{x \in G}\, x \in \Lambda$, $I(G)$ = augmentation ideal of $\mathbf{Z}G$. Each integer $r \in \mathbf{Z}$ prime to n determines a two-sided ideal (r, σ_G) in Λ. The fibre product

$$\begin{array}{ccc} \Lambda & \longrightarrow & \Lambda/I(G) \cong \mathbf{Z} \\ \downarrow & & \downarrow \\ \Lambda/(\sigma_G) & \longrightarrow & \mathbf{Z}/n\mathbf{Z} \end{array}$$

yields an exact sequence

$$u(\mathbf{Z}/n\mathbf{Z}) \xrightarrow{\delta} D(\Lambda) \longrightarrow D(\Lambda/(\sigma_G)) \longrightarrow 0.$$

The coset $r + n\mathbf{Z}$ maps onto the class of (r, σ_G) in $D(\Lambda)$. Let $T(G) = \operatorname{im} \delta$, a subgroup of $D(\Lambda)$. Ullom proves:

(a) A surjection $G \longrightarrow \bar{G}$ induces a surjection $T(G) \longrightarrow T(\bar{G})$.

(b) For H a subgroup of G, the restriction map gives a surjection $T(G) \longrightarrow T(H)$, and carries (r, σ_G) onto (r, σ_H).

(c) $T(G) = 0$ if G is cyclic.

(d) The exponent of $T(G)$ divides the Artin exponent of G.

(e) If G is an elementary abelian p-group of order p^{s+1}, where p is an odd prime, then $T(G)$ is cyclic of order p^s.

(f) For G metacyclic of order pq as in (7.7), $|T(G)| = q/(q, 2)$.

(g) For $G = S_n$ or A_n and p an odd prime, p divides $|T(G)|$ if and only if $p \leqslant n/2$.

(h) $T(G) \neq 1$ if G contains a noncyclic subgroup of odd order.

4. It does not seem likely that we can calculate $|D(\mathbf{Z}S_n)|$ and $|D(\mathbf{Z}A_n)|$ explicitly by these methods, except for small values of n.

9. Picard Groups

The main references for the material in this section are Bass [1], Fröhlich [7], Fröhlich, Reiner and Ullom [11], and MO, §§37, 38. We shall first define the Picard group Pic Λ of an arbitrary ring Λ (not necessarily commutative). To start with, let Λ, Δ be a pair of rings, and let ${}_\Lambda M_\Delta$ be a bimodule (that is, M is a left Λ-, right Δ-module such that $\lambda(m\delta) = (\lambda m)\delta$ for all λ, m, δ). We say that M is *invertible*, with inverse N, if there exists a bimodule ${}_\Delta N_\Lambda$ and bimodule isomorphisms

(9.1) $$M \otimes_\Delta N \cong \Lambda, \qquad N \otimes_\Lambda M \cong \Delta,$$

for which the following diagrams commute:

(9.2)
$$\begin{array}{ccc} M \otimes_\Delta N \otimes_\Lambda M & \longrightarrow & \Lambda \otimes_\Lambda M \\ \downarrow & & \downarrow \\ M \otimes_\Delta \Delta & \longrightarrow & M, \end{array} \qquad \begin{array}{ccc} N \otimes_\Lambda M \otimes_\Delta N & \longrightarrow & \Delta \otimes_\Delta N \\ \downarrow & & \downarrow \\ N \otimes_\Lambda \Lambda & \longrightarrow & N. \end{array}$$

In this case, we call the rings Λ, Δ *Morita equivalent.* There is then an equivalence of categories

$$_\Lambda\mathcal{M} \rightleftarrows {}_\Delta\mathcal{M}, \qquad _\Lambda X \to N \otimes_\Lambda X, \qquad _\Delta Y \to M \otimes_\Delta Y,$$

that is, there is a one-to-one correspondence between isomorphism classes of left Λ-modules X and of left Δ-modules Y, and this correspondence preserves Homs.

For example, if M is a free right Δ-module on n generators, and if we set

(9.4) $$\Lambda = \mathrm{Hom}_\Delta(M, M), \qquad N = \mathrm{Hom}_\Delta(M, \Delta),$$

then M is an invertible (Λ, Δ)-bimodule with inverse N. Further, in this case Λ is isomorphic to the matrix ring $M_n(\Delta)$. More generally, we shall call M_Δ a *progenerator* (for the category of right Δ-modules) if M is a finitely generated projective right Δ-module, such that Δ is a direct summand of $M^{(k)}$ for some k. Then we have (MO, (16.9))

(9.5) Proposition. *If M_Δ is a progenerator, and Λ, N are as in* (9.4), *then ${}_\Lambda M_\Delta$ is invertible with inverse N. Conversely, for any invertible bimodule ${}_\Lambda M_\Delta$ over a pair of rings Λ, Δ, the module M_Δ is a progenerator, and the formulas in* (9.4) *are isomorphisms* (*rather than equalities*).

Consider now the case where $\Delta = \Lambda$, and let (M) range over all isomorphism classes of invertible (Λ, Λ)-bimodules. These classes form a multiplicative group Pic Λ, the *Picard group* of Λ, where multiplication is given by $(M)(M') = (M \otimes_\Lambda M')$. The identity element is (Λ), and $(M)^{-1} = (N)$, with N as in (9.4). It is easily shown that if Λ and Γ are Morita equivalent, then Pic $\Lambda \cong$ Pic Γ (see MO, (37.9)).

Now let C denote the center of Λ, and consider those invertible (Λ, Λ)-bimodules M such that

$$cm = mc \quad \text{for all } m \in M,\ c \in C. \tag{9.6}$$

The classes (M) of such bimodules form a subgroup Picent Λ of Pic Λ. If Γ is Morita equivalent to Λ, then Picent $\Gamma \cong$ Picent Λ (see MO, (37.9)). We note also (MO, (37.21) and (37.22)):

(9.7) THEOREM. (i) Picent $A = 1$ *for any semisimple artinian ring* A.
(ii) Picent $\Lambda = 1$ *for any commutative semilocal*[11] *ring* Λ.

From now on, let Λ be an R-order in a semisimple F-algebra A, where char $F = 0$, and let C denote the center of Λ. We wish to make the bimodule isomorphism classes of two-sided Λ-ideals in A into a multiplicative group, but this cannot be done because of the absence of inverses. We must therefore restrict the types of ideals considered: an *invertible* Λ-ideal is a two-sided Λ-module $M \subset A$, such that $MN = NM = \Lambda$ for some two-sided Λ-module $N \subset A$. It follows at once that there are bimodule surjections

$$M \otimes_\Lambda N \longrightarrow \Lambda, \qquad N \otimes_\Lambda M \longrightarrow \Lambda,$$

and therefore (MO, (16.7)) the bimodule ${}_\Lambda M_\Lambda$ is invertible, with inverse N. We may remark that

$$N = \mathrm{Hom}_\Lambda({}_\Lambda M, {}_\Lambda \Lambda) \cong \{a \in A : Ma \subset \Lambda\}.$$

Each invertible Λ-ideal M obviously satisfies condition (9.6). Conversely, it turns out that every invertible (Λ, Λ)-bimodule, for which (9.6) holds, is bimodule isomorphic to an invertible Λ-ideal in A. Thus Picent Λ may be described as the group of classes of invertible Λ-ideals in A, and as such, is a natural generalization of the ideal class group Cl R. In fact, we obtain easily:

$$\text{Picent } \Lambda \cong I(\Lambda)/\{\Lambda c : c \in u(FC)\}, \tag{9.8}$$

where $I(\Lambda)$ is the group of invertible Λ-ideals in A, and FC is the center of A.

(The fact that the elements of Picent Λ are represented by Λ-ideals is a consequence of (9.7(i)). Condition (9.6) is needed in order to permit the use of the Skolem-Noether Theorem: Every automorphism of A which is the identity map on FC must be an inner automorphism.)

Every two-sided Λ-ideal in A is of course also a left Λ-ideal, and bimodule isomorphism implies isomorphism. Hence if F is an algebraic number field, it follows from the Jordan-Zassenhaus Theorem (1.7) that Picent Λ is finite.

In certain cases, we can compute Picent Λ directly from (9.8). For example, we have (MO, (37.27)).

(9.9) THEOREM. *Let A be a simple algebra whose center F is an algebraic number field, and let Λ be a maximal R-order in A. For a prime ideal P of R, let $A_P \cong M_n(D)$, where D is a skewfield, and set $m_P = (D : F_P)^{1/2}$, the* local index *of A at P. Then* Picent Λ_P

[11] See (2.2).

is a cyclic group of order m_P. *Furthermore,* $m_P = 1$ *for almost all P.*

OUTLINE OF PROOF. The group $I(\Lambda_P)$ is an infinite cyclic group generated by rad Λ_P. If π denotes a prime element of R_P, then $\pi\Lambda_P = (\text{rad } \Lambda_P)^{m_P}$. Since $FC = F_P$ in this case, the denominator on the right-hand side of (9.8) consists of all powers $\pi^r\Lambda_P$, $r \in \mathbf{Z}$. Therefore Picent $\Lambda_P \cong \mathbf{Z}/m_P\mathbf{Z}$, as asserted. The fact that $m_P = 1$ a.e. follows from an easy argument involving discriminants (see MO, (25.7)).

The calculation of Picent Λ_P, in case Λ_P is not a maximal order, is rather difficult. One approach involves the calculation of the automorphisms of Λ_P, as we now explain. Let Aut Λ denote the group of automorphisms of Λ, and In Λ the subgroup consisting of all inner automorphisms $x \longrightarrow uxu^{-1}$, $x \in \Lambda$, where u ranges over all units of Λ. We set

(9.10)
$$\begin{cases} \text{Autcent } \Lambda = \{f \in \text{Aut } \Lambda : f(c) = c \text{ for all } c \in C\}, \\ \text{Outcent } \Lambda = \text{Autcent } \Lambda/\text{In } \Lambda. \end{cases}$$

Now let M be an invertible Λ-ideal in A, and let $f \in$ Autcent Λ. We define a new bimodule M_f having the same elements as M, but with the action of Λ given by

$$\lambda_1 \circ m \circ \lambda_2 \ (\text{in } M_f) = \lambda_1 m f(\lambda_2) \ (\text{in } M).$$

For a pair of invertible Λ-ideals M, M', it turns out (see MO, (37.16)) that $M' \cong M$ as left Λ-modules if and only if $M' \cong M_f$ for some $f \in$ Autcent Λ. Furthermore (MO, (37.14)) there is a bimodule isomorphism $M \cong M_f$ if and only if $f \in$ In Λ. This readily implies that the map

(9.11) $$\omega : \text{Outcent } \Lambda \longrightarrow \text{Picent } \Lambda, \text{ given by } \omega(f) = (\Lambda_f),$$

is a monomorphism of groups.

In some cases, the map ω is an isomorphism. For example, suppose that $\Lambda_P = R_P G$ is an integral group ring. Now for each invertible Λ_P-ideal M we have $F_P M = A_P$, and M is projective as left Λ_P-module. Hence by a result of Swan (see CR, (77.14)) it follows that $M \cong \Lambda_P$ as left Λ_P-modules. Therefore when $\Lambda_P = R_P G$ we have Outcent $\Lambda_P \cong$ Picent Λ_P, and the calculation of Picent Λ_P reduces to that of Outcent Λ_P. This fact is used in Fröhlich's calculations [7] of Picard groups.

We are still faced with the question of determining Outcent Λ in practice. This can best be done by means of an alternate description of this group. Every $f \in$ Autcent Λ extends to an automorphism of A which fixes the center of A, and hence (Skolem-Noether Theorem, MO, §7d) is an inner automorphism. Now set

(9.12) $$\textit{normalizer} \text{ of } \Lambda = N(\Lambda) = \{a \in u(A) : a\Lambda a^{-1} = \Lambda\}.$$

The preceding remarks yield

(9.13) $$\text{Outcent } \Lambda \cong N(\Lambda)/u(\Lambda)\,u(FC),$$

with the isomorphism induced by mapping each $a \in N(\Lambda)$ onto the coset containing the automorphism $\lambda \longrightarrow a\lambda a^{-1}$, $\lambda \in \Lambda$.

The relation between global and local Picard groups is given by the following basic result of Fröhlich [7] (see MO, (37.28)):

(9.14) THEOREM. *Let Λ be an R-order in a semisimple F-algebra A, and let C be the center of Λ. Then there is an exact sequence*

$$1 \longrightarrow \text{Picent } C \xrightarrow{\tau} \text{Picent } \Lambda \longrightarrow \sum{}^{\oplus} \text{Picent } \Lambda_P \longrightarrow 1,$$

and Picent $\Lambda_P = 1$ *a. e.*

REMARKS. The map τ is given by $\tau(L) = (L\Lambda)$ for each invertible C-ideal L in FC. The fact that Picent $\Lambda_P = 1$ a.e. can be deduced from (9.9), as follows: let Λ' be a maximal R-order in A containing Λ. Then $\Lambda_P = \Lambda'_P$ a.e., and Picent $\Lambda'_P = 1$ a.e. by (9.9). Hence also Picent $\Lambda_P = 1$ a.e.

An invertible Λ-ideal M is called *locally free* if M is in the same genus as Λ as left Λ-module, that is, $M_P \cong \Lambda_P$ as left Λ_P-modules for each prime ideal P of R. As a matter of fact, any two-sided Λ-ideal which is locally free is automatically invertible, and is also locally free as right Λ-module. The bimodule isomorphism classes (M) of locally free invertible Λ-ideals M form a subgroup $LFP(\Lambda)$ of Picent Λ; this subgroup is called the *locally free Picard group* of Λ. There are several important cases in which $LFP(\Lambda)$ coincides with Picent Λ; this holds when Λ is commutative, by (9.7(ii)). It also holds when Λ is a maximal order. More important, by Swan's Theorem (1.8) we have $LFP(\mathbf{Z}G)$ = Picent $\mathbf{Z}G$ for each finite group G.

The remarks following (9.11) show that

$$LFP(\Lambda_P) \cong \text{Outcent } \Lambda_P \cong N(\Lambda_P)/u(\Lambda_P)u(F_PC).$$

Furthermore, (9.14) gives the exact sequence

$$1 \longrightarrow \text{Picent } C \longrightarrow LFP(\Lambda) \longrightarrow \sum{}^{\oplus} LFP(\Lambda_P) \longrightarrow 1.$$

These formulas permit the calculation of $LFP(\Lambda)$ in various cases. Let us mention a few of the results of Fröhlich [7]:

(i) If G is a p-group, then so is $\Sigma^{\oplus}$ Picent R_PG.

(ii) If G is a p-group, so is $D(C)$, where C is the center of $\mathbf{Z}G$.

(iii) If G is a dihedral group of order $2p$, with p an odd prime, then Picent $\mathbf{Z}G \cong D(C) \oplus \text{Cl } S$, where C is the center of $\mathbf{Z}G$, and S = alg. int. $\{Q(\omega + \omega^{-1})\}$, with $\omega = \sqrt[p]{1}$. Further, $D(C)$ is cyclic of order $(p-1)/2$.

(iv) For G the quaternion or dihedral group of order 8, we have Picent $\mathbf{Z}G \cong D(\mathbf{Z}G)$; hence Picent $\mathbf{Z}G$ is trivial for the dihedral case, and has order 2 for the quaternion case (see (6.4), (7.4)).

(v) Let I be an ideal of R, and let $A = M_n(F)$, $\Lambda = R \cdot 1 + I \cdot M_n(R)$, so Λ is a congruence order in A. Then

$$\text{Picent } \Lambda \cong \text{Cl } R \oplus PGL(n, R/I),$$

where

$$PGL(n, R/I) = GL(n, R/I)/u(R/I)$$

is the projective general linear group of $n \times n$ matrices over the ring R/I.

To conclude this section, we shall describe some results of Fröhlich, Reiner and Ullom [11] concerning the relation between the locally free Picard group $LFP(\Lambda)$ and the locally free class group Cl Λ of an order Λ.

(9.15) THEOREM. *There is a homomorphism*

$$\theta : LFP(\Lambda) \longrightarrow \mathrm{Cl}\,\Lambda,$$

given by $\theta(X) = [X]$ *for* $(X) \in LFP(\Lambda)$. *If the cancellation law holds for locally free* Λ*-ideals in* A, *then* $\ker\theta \cong \mathrm{Outcent}\,\Lambda$.

OUTLINE OF PROOF. Let X, Y be locally free invertible Λ-ideals in A, and take $X \subset \Lambda$ without loss of generality. Then there is an exact sequence of bimodules

$$0 \longrightarrow X \longrightarrow \Lambda \longrightarrow T \longrightarrow 0,$$

with T an R-torsion module. Since ${}_{\Lambda}Y$ is projective, we obtain an exact sequence

$$0 \longrightarrow X \otimes Y \longrightarrow Y \longrightarrow T \otimes Y \longrightarrow 0,$$

where $\otimes$ means $\otimes_{\Lambda}$. But $T \otimes Y \cong T$ since Y is locally free. Comparing these two sequences and using Schanuel's Lemma, we obtain a left Λ-isomorphism $X \dotplus Y \cong \Lambda \dotplus X \otimes Y$. Further, $X \otimes Y \cong XY$, and therefore $[X] + [Y] = [XY]$ in $\mathrm{Cl}\,\Lambda$. This proves that θ is a homomorphism. Finally, suppose that $\theta(X) = 0$, so $[X] = 0$ in $\mathrm{Cl}\,\Lambda$. Assuming cancellation, we get $X \cong \Lambda$ as left Λ-modules, whence $(X) = (\Lambda_f)$ for some $f \in \mathrm{Autcent}\,\Lambda$ by the remarks before (9.11). Thus $\ker\theta = \mathrm{im}\,\omega$, as claimed.

Suppose that Λ is commutative. We have already remarked that $LFP(\Lambda) = \mathrm{Picent}\,\Lambda$ in this case. Further, it is clear that the map θ in (9.15) is a surjection. Thus we obtain $\mathrm{Picent}\,\Lambda = LFP(\Lambda) \cong \mathrm{Cl}\,\Lambda$ whenever Λ is commutative. Thus, in this case the method of "adding" locally free ideals used to define the class group $\mathrm{Cl}\,\Lambda$ in §1 turns out to be equivalent (up to isomorphism) to the method of multiplying locally free ideals when defining the Picard group $\mathrm{Picent}\,\Lambda$.

Let us next obtain an explicit formula for $\operatorname{cok}\theta$ in terms of ideles. We define the *idele normalizer* of Λ by

$$\bar{N}(\Lambda) = \left\{ (x_p) \in \prod N(\Lambda_p) : x_p \in u(\Lambda_p) \text{ a.e.} \right\}.$$

Each $(X) \in LFP(\Lambda)$ is then expressible as $X = A \cap \{\bigcap \Lambda_p x_p\}$ for some $(x_p) \in \bar{N}(\Lambda)$. This gives a surjection of $\bar{N}(\Lambda)$ onto $LFP(\Lambda)$, and yields an isomorphism

$$\phi^* : \bar{N}(\Lambda)/u(\Lambda)u(FC) \cong LFP(\Lambda).$$

On the other hand, each $x \in \bar{N}(\Lambda)$ determines an element $\theta^*(x) \in JK(A)$. The homomorphism θ^* induces a map

$$\theta^* : \bar{N}(\Lambda)/u(\Lambda)u(FC) \longrightarrow JK(A)/\mathrm{im}\,K_1(A) \cdot \mathrm{im}\,UK(\Lambda),$$

and we obtain a commutative diagram

$$\begin{array}{ccc} \bar{N}(\Lambda)/u(\Lambda)u(FC) & \xrightarrow{\ \theta^*\ } & JK(A)/\mathrm{im}\,K_1(A) \cdot \mathrm{im}\,UK(\Lambda) \\ \Big\downarrow{\scriptstyle \phi^*} & & \Big\downarrow{\scriptstyle \phi} \\ LFP(\Lambda) & \xrightarrow[\ \theta\ ]{} & \mathrm{Cl}\,\Lambda. \end{array}$$

Therefore $\operatorname{cok}\theta \cong \operatorname{cok}\theta^*$.

For convenience, we restrict our attention to the case where F is an algebraic number field. We may then use the formula for $\mathrm{Cl}\,\Lambda$ given in (2.21), and calculate $\operatorname{cok}\theta^*$ as a quotient of $I(FC)/(FC)^{+} \cdot \prod nr\, u(\Lambda_p)$. We obtain

(9.16) THEOREM. *Let us put*

$$nr\,\bar{N}(\Lambda) = \{(nr\,x_P) \in J(FC) : (x_P) \in \bar{N}(\Lambda)\}.$$

Then

$$\operatorname{cok}\theta \cong J(FC)/(FC)^{+} \cdot nr\,\bar{N}(\Lambda) \tag{9.17}$$

if F is an algebraic number field.

Formula (9.17) may be restated in terms of ideals, rather than ideles; for the explicit statement, see [11, Theorem 5.13]. An interesting consequence of the ideal-theoretic version is as follows:

(9.18) COROLLARY. *Keeping the notation of* (2.17), *let* $(A_i : F_i) = m_i^2$, $1 \leqslant i \leqslant s$. *Then the exponent of* $\operatorname{cok}\theta$ *divides the least common multiple of* $m_1, \ldots, m_s$.

Finally, assume that stable isomorphism implies isomorphism for locally free Λ-lattices, and let $\Gamma = M_n(R) \otimes_R \Lambda \cong M_n(\Lambda)$. Then there is a commutative diagram with exact rows

$$\begin{array}{ccccccc} 1 & \longrightarrow & \text{Outcent}\,\Lambda & \longrightarrow & LFP(\Lambda) & \longrightarrow & \text{Cl}\,\Lambda \\ & & \downarrow{\scriptstyle\alpha} & & \downarrow{\scriptstyle\beta} & & \downarrow{\scriptstyle\gamma} \\ 1 & \longrightarrow & \text{Outcent}\,\Gamma & \longrightarrow & LFP(\Gamma) & \longrightarrow & \text{Cl}\,\Gamma, \end{array}$$

where $\alpha(f) = 1 \otimes f$, and β, γ are "change of rings" maps. As shown in [11],

$$\ker\gamma = \{x \in \text{Cl}\,\Lambda : nx = 0\}, \quad \operatorname{cok}\gamma \cong \text{Cl}\,\Lambda/n\cdot\text{Cl}\,\Lambda.$$

In particular, if Λ is a commutative R-order, it follows that

$$\text{Outcent}\,M_n(\Lambda) \cong \{x \in \text{Cl}\,\Lambda : nx = 0\}.$$

Thus

$$\text{Outcent}\,M_n(R) \cong \{x \in \text{Cl}\,R : nx = 0\}.$$

Therefore for each automorphism f of $M_n(R)$ fixing its center, f^n is an inner automorphism of $M_n(R)$; this result was first proved by Rosenberg and Zelinsky [36a].

The list of references which follows includes almost all articles on class groups and Picard groups which have appeared up to this time, or are available in preprint form. It also includes a number of basic texts for further reading about the various topics considered in these lectures. These texts, arranged by subject matter, are:

Homological algebra: [14], [37],

Algebraic K-theory: [1], [20], [23], [40a], [41],

Representation theory: [3], [5a], [14a],

Integral representations: [3], [26], [34], [35].

REFERENCES

1. H. Bass, *Algebraic K-theory*, Benjamin, New York, 1968. MR **40** #2736.

2. P. Cassou-Noguès, *Classes d'idéaux de l'algèbre d'un groupe abélian,* C. R. Acad. Sci. Paris Sér. A-B **276** (1973), A973–A975. MR **47** #8667.

3. C. W. Curtis and I. Reiner, *Representation theory of finite groups and associative algebras,* Pure and Appl. Math., vol. XI, Interscience, New York, 1962; 2nd ed., 1966. MR **26** #2519.

4. S. Endo and T. Miyata, *Quasi-permutation modules over finite groups.* I, II, J. Math. Soc. Japan **25** (1973), 397–421; **26** (1974), 698–713. MR **47** #6823.

5. ———, *On the projective class group of finite groups* (to appear).

5a. W. Feit, *Characters of finite groups,* Math. Lecture Notes Series, Benjamin, New York, 1967. MR **36** #2715.

6. A. Fröhlich, *On the classgroup of integral group rings of finite abelian groups.* I, II, Mathematika **16** (1969), 143–152; ibid. **19** (1972), 51–56. MR **41** # 5512; **48** #394.

7. ———, *The Picard group of noncommutative rings, in particular of orders,* Trans. Amer. Math. Soc. **180** (1973), 1–45. MR **47** #6751.

8. ———, *Locally free modules over arithmetic orders,* J. Reine Angew. Math. **274/275** (1975), 112–124.

9. ———, *Module invariants and root numbers for quaternion fields of order* $4l^r$, Proc. Cambrdige Philos. Soc. **76** (1974), 393–399.

10. A. Fröhlich, M. E. Keating and S. M. J. Wilson, *The class group of quaternion and dihedral* 2*-groups,* Mathematika **21** (1974), 64–71.

11. A. Fröhlich, I. Reiner and S. Ullom, *Picard groups and class groups of orders,* Proc. London Math. Soc. (3) **29** (1974), 405–434.

12. S. Galovich, *The class group of a cyclic p-group,* J. Algebra **30** (1974), 368–387.

13. S. Galovich, I. Reiner and S. Ullom, *Class groups for integral representations of metacyclic groups,* Mathematika **19** (1972), 105–111. MR **48** #4087.

14. P. J. Hilton and U. Stammbach, *A course in homological algebra,* Graduate Texts in Math., vol. 4, Springer-Verlag, Berlin and New York, 1971. MR **49** #10751.

14a. B. Huppert, Endliche Gruppen. I, Die Grundlehren der math. Wissenschaften, Band 134, Springer-Verlag, Berlin and New York, 1967. MR **37** #302.

15. H. Jacobinski, *Über die Geschlechter von Gittern über Ordnungen,* J. Reine Angew. Math. **230** (1968), 29–39. MR **37** #5250.

16. ———, *Genera and decompositions of lattices over orders,* Acta. Math. **121** (1968), 1–29. MR **40** #4294.

17. H. Jacobinski, *Two remarks about hereditary orders,* Proc. Amer. Math. Soc. **28** (1971), 1–8. MR **42** #7688.

18. M. E. Keating, *Class groups of metacyclic groups of order $p^r q$, p a regular prime,* Mathematika **21** (1974), 90–95.

19. M. A. Kervaire and M. P. Murthy, *On the projective class group of cyclic groups of prime power order* (to appear).

20. T. Y. Lam and M. K. Siu, *K_0 and K_1–an introduction to algebraic K-theory,* Amer. Math. Monthly **82** (1975), 329–364.

21. M. P. Lee, *Integral representations of dihedral groups of order 2p*, Trans. Amer. Math. Soc. **110** (1964), 213–231. MR **28** #139.

22. J. Martinet, *Modules sur l'algèbre du groupe quaternionien,* Ann. Sci. École Norm. Sup. (4) **4** (1971), 399–408. MR **45** #302.

22a. A. Matchett, *Bimodule-induced morphisms of class groups* (to appear).

23. J. Milnor, *Introduction to algebraic K-theory,* Ann. of Math. Studies, Princeton Univ. Press, Princeton, N. J., 1971.

23a. T. Nakayama and Y. Matsushima, *Über die multiplikative Gruppe einer p-adischen Divisionsalgebra,* Proc. Imp. Acad. Tokyo **19** (1943), 622–628. MR **7**, 238.

24. L. C. Pu, *Integral representations of non-abelian groups of order pq,* Michigan Math. J. **12** (1965), 231–246. MR **31** #2321.

25. I. Reiner, *A survey of integral representation theory,* Bull. Amer. Math. Soc. **76** (1970), 159–227. MR **40** #7302.

26. ———, *Maximal orders,* Academic Press, London, 1975.

27. ———, *Hereditary orders,* Rend. Sem. Mat. Univ. Padova **52** (1974), 219–225.

27a. ———, *Locally free class groups of orders.* Carleton University Lecture Notes **9** (1974), 21.01–21.29 (Springer Lecture Notes, vol. 488).

28. ———, *Projective class groups of symmetric and alternating groups,* Linear and Multilinear Algebra **3** (1975), 115–121.

29. I. Reiner and S. Ullom, *Class groups of integral group rings,* Trans. Amer. Math. Soc. **170** (1972), 1–30. MR **46** #3605.

30. ———, *A Mayer-Vietoris sequence for class groups,* J. Algebra **31** (1974), 305–342.

31. ———, *Class groups of orders, and a Mayer-Vietoris sequence,* Lecture Notes in Math., vol. 353, Springer-Verlag, Berlin and New York, 1973, pp. 139–151. MR **50** #457.

32. ———, *Remarks on class groups of integral group rings,* Symp. Math. Ist. Nazionale Alta Mat. (Rome) **13** (1974), 501–516.

33. D. S. Rim, *Modules over finite groups,* Ann. of Math. (2) **69** (1959), 700–712. MR **21** #3474.

34. K. W. Roggenkamp, *Lattices over orders.* II, Lecture Notes in Math., vol. 142, Springer-Verlag, Berlin and New York, 1970. MR **44** #247b.

35. K. W. Roggenkamp and V. Huber-Dyson, *Lattices over orders.* I, Lecture Notes in Math., vol. 115, Springer-Verlag, Berlin and New York, 1970. MR **44** #247a.

36. M. Rosen, *Representations of twisted group rings,* Ph. D. Thesis, Princeton Univ., 1963.

36a. A. Rosenberg and D. Zelinsky, *Automorphisms of separable algebras,* Pacific J. Math. **11** (1961), 1109–1117. MR **26** #6215.

37. J. J. Rotman, *Notes on homological algebra,* van Nostrand Reinhold, New York, 1970.

38. R. G. Swan, *Induced representations and projective modules,* Ann. of Math. (2) **71** (1960), 552–578. MR **25** #2131.

39. ———, *Projective modules over group rings and maximal orders,* Ann. of Math. (2) **76** (1962), 55–61. MR **25** #3066.

40. ———, *The Grothendieck ring of a finite group,* Topology **2** (1963), 85–110. MR **27** #3683.

40a. ———, *Algebraic K-theory,* Lecture Notes in Math., vol. 76, Springer-Verlag, Berlin and New York, 1968. MR **39** #6940.

41. R. G. Swan and E. G. Evans, *K-theory of finite groups and orders,* Lecture Notes in Math., vol. 149, Springer-Verlag, Berlin and New York, 1970. MR **46** #7310.

42. S. Ullom, *A note on the classgroup of integral group rings of some cyclic groups,* Mathematika **17** (1970), 79–81. MR **42** #4650.

43. ———, *The exponent of class groups,* J. Algebra **29** (1974), 124–132. MR **49** #2910.

43a. ———, *Nontrivial lower bounds for class groups of integral group rings* (to appear).

43b. ———, *The Δ-decomposition of the class group of cyclic p-groups* (to appear).

44. S. M. J. Wilson, *Reduced norms in the K-theory of orders,* J. Algebra (to appear).

45. C. T. C. Wall, *On the classification of hermitian forms.* I–V, Compositio Math. **22** (1970), 425–451; Invent. Math. **18** (1972), 119–141; ibid. **19** (1973), 59–71; ibid. **23** (1974), 241–260; ibid. **23** (1974), 261–288. MR **43** #7425; **48** #2184; #2185.

46. ———, *Norms of units in group rings,* Proc. London Math. Soc. (3) **29** (1974), 593–632.

Index